Sheffield Hallam University
Learning and IT Services
Adsetts Centre City Campus

D1423729

THIRD EDITION

DICTIONARY OF

STATISTICS & METHODOLOGY

SHEFFIELD HALLAM UNIVERSITY
LEARNING CENTRE
WITHDRAWN FROM STOCK

THIRD EDITION

DICTIONARY OF
STATISTICS &
METHODOLOGY

A NONTECHNICAL GUIDE
FOR THE SOCIAL SCIENCES

W. PAUL VOGT
Illinois State University

SAGE Publications
Thousand Oaks ■ London ■ New Delhi

Copyright © 2005 by Sage Publications, Inc.

All rights reserved. No part of this book may be reproduced or utilized in any form or by any means, electronic or mechanical, including photocopying, recording, or by any information storage and retrieval system, without permission in writing from the publisher.

For information:

Sage Publications, Inc.
2455 Teller Road
Thousand Oaks, California 91320
E-mail: order@sagepub.com

Sage Publications Ltd.
1 Oliver's Yard
55 City Road
London EC1Y 1SP
United Kingdom

Sage Publications India Pvt. Ltd.
B-42, Panchsheel Enclave
Post Box 4109
New Delhi 110 017 India

Printed in the United States of America

Library of Congress Cataloging-in-Publication Data

Vogt, W. Paul.
Dictionary of statistics & methodology: A nontechnical
guide for the social sciences / W. Paul Vogt.—3rd ed.
    p. cm.
Includes bibliographical references.
ISBN 0-7619-8854-8 (cloth)—ISBN 0-7619-8855-6 (pbk.)
    1. Social sciences—Statistical methods—Dictionaries. 2. Social
sciences—Methodology—Dictionaries. I. Title.
HA17.V64 2005
300.′ 1′5195—dc22

2004027624

05   06   07   08   09   10   9   8   7   6   5   4   3   2   1

Acquisitions Editor:	Lisa Cuevas Shaw
Editorial Assistant:	Karen Gia Wong
Production Editor:	Melanie Birdsall
Copy Editor:	Bill Bowers, Interactive Composition Corporation
Typesetter:	C&M Digitals (P) Ltd.
Proofreader:	Teresa Herlinger
Cover Designer:	Glenn Vogel

SHEFFIELD HALLAM UNIVERSITY
519.50243
VO

Contents

List of Figures

List of Tables

Preface to the Third Edition

Audience

Since the original edition of this dictionary was published in 1993, a few other dictionaries have appeared (see the Suggestions for Further Reading). While each of these volumes has its strengths, the book you have in front of you is the most useful for its intended audience. That audience is composed of readers trying to inform themselves about research terms and concepts so that they can be effective consumers of research. That general audience is a broad group. It ranges from professionals who want to improve their practice by keeping up with the research in their fields to students trying to decipher research reports for the first time. Another group that frequently consults this dictionary is made up of students planning eventually to conduct their own research, but who need some help reviewing the research literature before they begin.

Features

Without going into detailed comparisons with other volumes, I will briefly list here what I see as this dictionary's noteworthy features.

- *Nontechnical.* By being nontechnical, this work adheres to one of the most basic rules of good dictionary writing: a definition should never contain terms more obscure than the one being defined. For example, definition by algebraic formula (very common in other dictionaries) is helpful for mathematicians, but is of limited use for less specialized readers.
- *Self-contained.* Each term used in the definitions is itself defined in the dictionary. When other technical terms are used in a definition, they are cross-referenced. Readers with a good grasp of ordinary English will rarely if ever come to a dead end because they have looked up a word only to see it defined by a word they do not know and that they can look up only by consulting another volume.
- *Descriptive not prescriptive.* Terms are described as they are actually used rather than how they "should" be used. Setting standards for proper

use is important, but a dictionary is more valuable to readers when it is informative rather than normative.

- *Comprehensive.* My goal is that this book should include everything that readers are likely to need to understand a research report. I have tried to make the book comprehensive in several ways.

  - Methodology and design terms are defined, as well as terms relating to statistical analysis and measurement.
  - Concepts from qualitative research methods are covered (although less extensively) in addition to terms from the more quantitative traditions in research.
  - Elementary terms and concepts are defined, often extensively, not merely more specialized and advanced terms.
  - Related terms from theory (e.g., postmodernism, positivism, constructionism) and philosophy (e.g., empiricism, nominalism, realism) are defined. The rationale for these inclusions is, as always, that readers are likely to encounter such terms when reading research reports.

New in the Third Edition

The main difference between this edition and the previous two is that this one is longer, reflecting growth in the field. The number of definitions and illustrations has grown from about 2,000 in the second edition to about 2,400 in this one, an increase of around 20%. While I have done some trimming to make room for the new, comparatively few terms from the earlier editions have been deleted. Old terms linger on; new terms are invented much faster than old ones die out.

Sources

Where did the new definitions come from? Two main sources provided the new entries in this edition. It is a long-standing practice of dictionary writers to comb the relevant literature for terms and their use. As Dr. Johnson said of his famous 18th-century dictionary, I decided what to include by the "perusal of our writers." Since the publication of the second edition of this dictionary, I have tried to be quite systematic in my efforts to review relevant literature in the social sciences, keeping alert to the appearance of new methods, new terms, and newly popular terms. But this book is not an unabridged dictionary, in which *any* use would justify inclusion. Rather, I have had to exercise judgment and include only fairly widespread terms.

The perusal of our methodology writers made clear the need for new terms when new methods were invented. Just as frequently, new labels for existing methods have come into use. It is remarkably common in quantitative research to find that there are several terms used to describe the same concept or technique. This proliferation of synonyms is perhaps even more true in the field of qualitative methods, which is going through a period of rapid growth that has not yet been codified.

Queries and suggestions by readers have been my other chief source. Communications from readers of earlier editions have led me to include items I had omitted and to revise others. As always, I greatly appreciate this help from people I think of as volunteer co-editors. I look forward to receiving e-mails with more suggestions for improvement: wpvogt@ilstu.edu.

Preface to the Second Edition

M y aim for this second edition is the same as in the first: to give readers of research in the social and behavioral sciences an easy-to-use reference work for statistical concepts and methodological terms. A related goal is to enable people to be informed consumers and critical readers of research works that might otherwise be inaccessible to them. As always, every definition is "nontechnical." This means that definitions are written in ordinary English whenever possible, and they contain no formulas.

I have continued defining elementary concepts more fully than advanced concepts and highly specialized terms. I do that because it would be impossible, no matter how long the volume, to explain advanced topics in great detail. In this dictionary, definitions of advanced concepts tend to be short and use terms that are likely to be unfamiliar to readers with elementary knowledge of statistics and methodology. But, by following the paths suggested in the cross-references in the definitions of advanced terms, readers should be able to "work backward" until they reach definitions they understand. Conversely, the cross-references also suggest ways readers can "work forward" from elementary terms to more specialized topics.

Cross-references are also helpful for the reader with limited background, for another reason. As with any other "language," methodological and statistical discourse contains many synonyms (and even some homonyms). This means that one technique can be referred to in more than one way, sometimes even by the same author. I have paid extra attention in this edition to reducing that source of confusion.

The most visible change in this second edition is that it is substantially longer than the first, despite the fact that I have streamlined much of the original text. One virtue of a reference work is completeness. Another is portability. Unfortunately, these two lead an author in opposite directions, and a compromise must be made. In this edition, while completeness has had its say, I have tried to be concise while not forgetting the overriding need to make the book reasonably comprehensive. The first edition contained roughly 1,400 definitions and illustrations. This second edition includes about 2,000, or around one third more. Besides these additions, I have also made hundreds of smaller, less visible revisions. My goal as I made these changes was always to render the original definitions clearer, whether I was pruning, grafting, or simply rewriting.

In short, while the second edition follows the same plan as the first, it is, I believe, a considerable improvement. Since the response to the first edition was quite positive, I am hoping that this edition, too, proves useful to many explorers in the social and behavioral sciences and related fields such as education, management, business, and social work.

Preface to the First Edition

A lmost all research studies contain technical terms that nonspecialists do not know and, a much greater problem, that they cannot easily look up. What's needed to lower the jargon barrier between readers and research is a handy reference work where students and others can find quick definitions of a wide variety of statistical and methodological terms. I've tried to satisfy that need in this dictionary.

Introduction

This dictionary gives nontechnical definitions of statistical and methodological terms used in the social and behavioral sciences. Special attention is paid to terms that most often prevent educated general readers from understanding journal articles and books in sociology, psychology, economics, and political science—and in applied fields that build on those disciplines, such as education, policy studies, business, and administrative science. The dictionary provides definitions that will enable readers to get through a difficult article or passage. But it does not, for the most part, directly explain how to do research or how to compute the statistics briefly described.

The emphasis throughout is thus much more on concepts than calculations. Because the concepts are often complicated, readers may find that a definition makes sense only after it has been illustrated by an example. That is why the examples are frequent and sometimes longer than the definitions.

Quite a few terms are included that might not meet some strict definitions of "methodological" or "statistical"—for instance, "learning curve," "opportunity costs," "SPSS," and "zero-sum game." But they, and several others like them, are defined because they meet the main criteria for inclusion: They pop up fairly often in discussions related to research methods, they occur in more than one discipline, and many people are unsure about what they mean.

Verbal definitions are used to the exclusion of algebraic definitions, even when mathematical symbols and formulae would be more efficient—at least for users already familiar with statistics and social science methods. The general approach has been to treat the early stages of learning statistics more like studying a language than like learning a branch of mathematics.

When learning any language, beginners will sometimes be frustrated because they have to look up words in the definition of the term they just looked up. By writing the definitions in ordinary English whenever possible, I have tried to keep this unavoidable annoyance to a minimum. But it is unavoidable, and it is harder to avoid in a dictionary of technical terms than in a dictionary of a natural language—and there is no escape when defining advanced concepts that are built upon several more basic concepts. This is why many of the terms in this dictionary use other methodological or statistical terms in their definitions. Those terms, also defined in the dictionary, are indicated by an asterisk (*).

As in any language, in statistics and methodology, more than one word may be used to express the same idea. In such cases, I have defined fully what I believe to be the more common term and briefly defined and cross-referenced the others. But, I have not tried to stipulate the "proper" labels for concepts that appear under more than one name; nor have I specified the "correct" use of terms that are used in different ways. In short, I have attempted to be inclusive and descriptive, not exclusive and prescriptive. The goal throughout has been to provide a comprehensive dictionary of terms that will increase access to works in the social and behavioral sciences.

How to Use This Dictionary

1. When an entry contains other terms defined in this dictionary, they are indicated by an asterisk (*), especially those terms most likely to be helpful for reading that particular entry.

2. Entries are in alphabetical order, using the letter-by-letter (not word-by-word) method. This means that when looking up terms and expressions made up of more than one word, you should ignore the spaces and hyphens between words. For example, "F," "F Distribution," and "F Ratio" are separated by several pages, not grouped together as they would be using the word-by-word method. The only exception to the letter-by-letter rule is entries with a comma, such as "Association, Measure of."

3. If a term has more than one meaning, definitions are separated by letters—(a), (b), and so on—with the more common definitions usually coming first.

4. In terms containing numbers, the numbers are spelled out and alphabetized accordingly. For example, to find "2 × 2 design," look under "Two-by-Two Design."

5. Greek letters are anglicized, spelled out, and alphabetized accordingly. Consult the table "The Greek Alphabet" on the inside front cover for equivalents. For example, to find α, you would look under "Alpha."

6. For other symbols, consult the table "Frequently Used Symbols" on the inside front cover, get the equivalent in words, and look up the words in the ordinary way. For example, to find | |, you would look in the table to learn that it means "absolute value" and then consult that entry in the alphabetical listing.

A-B-A-B Designs Research designs that alternate *baseline measures of a *variable with measures of that variable after a *treatment. Such designs are used in single-subject or single-group *experiments having one treatment and no *control group. The "A" represents the baseline, the "B" represents the treatment. A-B and A-B-A designs are shorter versions of the same research procedure. A-B-A-B designs are often used when it would be unethical to withhold a treatment from a control group. See *interrupted time-series design.

For example, in order to see whether a treatment had an effect, patients' symptoms might be measured daily for four weeks. During the first week there would be no treatment (baseline, A). Then the treatment (B) would be given for a week. In the third week, the treatment could be withdrawn, and a second baseline (A) would be established. In the fourth week the treatment would be given again.

Abduction A term used to refer to the process of moving from everyday descriptions of the social world to social scientific descriptions and analyses of it.

Abscissa (a) The horizontal axis (or *x* axis) on a graph. (b) A particular point or value on that axis.

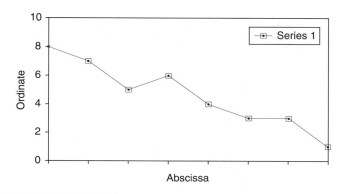

Figure A.1 Abscissa

Absolute Frequency   The actual number of times a particular value occurs in a *distribution, as opposed to its *relative frequency or the *proportion of times it occurs.

Absolute Deviation   See *mean absolute deviation, *average deviation.

Absolute Value (of a Number)   The value of a number regardless of its sign (positive or negative). Also called the "modulus."

For example, the absolute value of +8 and −8 is the same, which is 8. Symbolized |8|. (The absolute value is the positive value, since a number without a sign is positive.)

Absolute Zero Point   A value on a scale indicating that none of the variables being measured is present. See *ratio scale.

Abstraction   A cognitive process in which common elements of diverse things are identified and pulled out of their context (abstracted) in order to study them. Doing this makes it possible to classify and analyze *data that could not otherwise be studied together. Abstraction increases a researcher's ability to generalize, but at the cost of some of the empirical details of each case.

For example, say you wanted to do a comparative study of the relation of democracy to economic equality in modern nations. You might construct a "Democracy Index" (made up of some measures of freedom of the press, open elections, independent judiciary, etc.) and an "Equality Index" (made up of measures of the distribution of income, wealth, employment opportunities, etc.). You could then rank the nations using the two indexes and compute *correlation coefficients between nations' ranks on one index and their ranks on the other. On the basis of this work, you might be able to draw conclusions about whether there was a general relationship between economic equality and democracy.

Only by using such abstractions as your Democracy and Equality Indexes can you work at a *level of generality high enough to make comparisons. To the extent that you focused on the specific details of the economic and political situations in each country, it would be more difficult to make comparisons. And, conversely, the more you tried to make comparisons, the more difficult it would be to understand the concrete realities of economic and political life in each nation.

Accelerated Longitudinal Design   A combination of the methods (and their advantages and disadvantages) of *cross-sectional and *longitudinal studies in research on individuals over time. See *age effects, *stratified sample, *cohort sequential design.

For example, say you want to study the effects of aging on some variable(s) over the 30 years from age 20 to age 50. You do not have 30 years to conduct your study, so you cannot use a true longitudinal design and follow a group for 30 years (from age 20 to 50). You know that it is not justifiable simply to do a cross-sectional sample of subjects of different ages and assume that younger subjects will become like older subjects as they age. So you combine the two methods in an accelerated longitudinal design and take samples of subjects aged 20, 25, 30, 35, 40, and 45 and follow each group for five years.

Acceptance Error   Another term for *beta error, or *Type II error.

Acceptance Region   The range of values in a sample that would lead to the acceptance of the *null hypothesis. The opposite of the *regions of rejection.

Acceptance Sampling   A process of sampling designed to accept or reject a product.

   For example, an inspector might examine 1% of items manufactured. On the basis of the number of defects in that sample, the inspector can infer what the rate will be in the entire production run (*population).

Accidental Error   Error in measurement that cannot be predicted or controlled because the researcher is ignorant of its cause(s). Compare *bias, *random error.

Accidental Sample   A sample gathered haphazardly, for example, by interviewing the first 100 people you ran into on the street who were willing to talk to you. An accidental sample is not a *random sample. The main disadvantage of an accidental sample is that the researcher has no way of knowing what the *population might be. See *convenience sample, *probability sample.

Accurate to *n* Decimal Places (or Digits)   A statement that indicates how *rounding is done.

   For example, since $e = 2.71828 \ldots$, the approximation 2.7183 (with the 3 the result of rounding) is accurate to five decimal places or significant digits.

Acquiescent Response Style   The supposed tendency of some subjects, quite apart from what they actually believe, to be agreeable and say "yes" to all statements in an interview or on a questionnaire. Also called acquiescence bias. Compare *social desirability bias.

   There is not much evidence that such yea-saying is very common, no more common, at any rate, than a naysaying "quarrelsome response style." Also, as a matter of routine good practice, most question writers now compose their *scales and *indexes in such a way that persons would have to answer questions in opposite ways to convey their beliefs—for example, asking for a yes or no response to both statements "abortion should be legal" and "abortion should be illegal."

Action Research   A type of *applied research designed to find the most effective way to bring about a desired social change or solve a practical problem, usually in collaboration with the subjects of the research. Compare *basic research, *evaluation research.

Active Control Trial   A *clinical trial in which a *treatment is compared to another treatment rather than to a *placebo.

Active Variable   An experimental or manipulated *independent variable, as opposed to one over which the researcher exerts no control. Compare *attribute variable, *background variable.

Actuarial Statistics   Statistics used to calculate insurance rates and pensions. They estimate *risk.

Actuary   A statistician specializing in *probability theory, particularly as it relates to calculating risks for insurance companies. Actuaries were historically among the earliest professional statisticians. They estimate the likely dividends companies will have to pay, and therefore how high the premiums will have to be in order for them to make a profit.

AD   Abbreviation for *average deviation. See *mean absolute deviation (MAD).

Added Variable Plot   A graphic technique used to decide whether an additional *predictor (*independent) variable needs to be added to a *regression equation.

Additive   Said of a relation or a *model in which the effects of the *independent variables on a *dependent variable can simply be added together to find their total effect. Contrast *interaction effect.

Ad Hoc   Literally, Latin for "for this." Said of an explanation or a solution improvised for a specific purpose. Compare *planned comparisons.

Ad Infinitum   Latin for "continuing without end." Compare *limit.

Adjusted R^2   An *R^2 (R-squared) adjusted to give a truer (smaller) estimate of the degree to which the *independent variables in a *regression analysis explain the *dependent variable. The adjustment is made by taking into account the number of independent variables and the sample size. The adjusted R^2 is a measure of *strength of association. Compare *coefficient of determination, *omega squared.

Adjusting for Baseline   Any technique in longitudinal studies that takes the starting point into account when reporting on the end point. For example, in studies of academic achievement, students' test scores at the end of the year might be compared to their scores at the beginning of the year in order to measure improvement. This can be done by using *difference scores or by *controlling for the earlier scores. See *value added models.

Adjustment Statistically removing the effects of a variable in which the researcher is not interested in order to better focus on variables that are of interest. Various statistical techniques can be used to make this adjustment. For example, a researcher who wanted to study the effects of ethnicity on income would probably want to adjust for other variables that could account for any association, such as education level. See *control for, *crossbreaks, *regression analysis.

Admissible Hypothesis An hypothesis that is open to consideration, because it is not mathematically or logically self-contradictory or otherwise impossible.

Age Effects Outcomes attributable to subjects' ages, as when adults in their sixties have different values than those in their thirties. Compare *maturation effects.

Age effects are often difficult to distinguish from *period effects (people in the 1960s had different values than those in the 1990s) and *cohort effects, since people in their thirties in the 1960s were in their sixties in the 1990s!

Aggregate A group of persons, or other *units of analysis, that have certain traits or characteristics in common without necessarily having any direct social connection with one another, such as the population of a city. Also called *aggregation. Compare *holism, *disaggregate.

For example, "all female physicians" is an aggregate; so is "all European cities with populations over 20,000."

Gross National Income is an aggregation of data about individual incomes.

Aggregate Data Information about *aggregates or groups such as races, social classes, or nations. Sometimes contrasted with *micro-data. See *level of analysis.

Aggregation (a) A summary statistic composed of data about individual cases or units, such as the mean score on a test calculated from individuals' test scores. (b) The process of collecting data at a group level because they are not available at the individual level. See *ecological fallacy.

Aggregation Problem The difficulty of predicting macro-level behavior from micro-level data. The term is used mainly by economists, but the problem pervades the social sciences. See *ecological fallacy, *level of analysis, *abstraction.

A

Agreement, Coefficient of   (a) A measure of the relation of a single item on a *scale or an *index with the rest of a scale or index. See *Cronbach's alpha. (b) A measure of the similarity of one rating to another. See *inter-rater reliability.

Agreement, Measure of   Any of several statistics used to summarize the degree of agreement between the rankings or classifications made by two or more observers. Examples include *Cohen's kappa and *Kendall's coefficient of concordance.

Ahistorical   Said of an interpretation, theory, or point of view that ignores the influence of the past on the present, or that assumes the past is no different than the present, or that ignores time as a variable.

AI   *Artificial Intelligence.

Aleatory   Relating to chance, luck, randomness, or uncertainty. Sometimes, particularly in previous centuries, used in *probability theory; for example, "aleatory variable" means *random variable. More recently used in *qualitative research to mean uncaused or random.

Algorithm   (a) A set of clearly defined rules for solving a problem in a limited number of steps. In an ideal algorithm, if the rules are followed, the solution will require no judgment because it will automatically be correct. (b) Elements of or routines in a *computer program. Different statistical packages may use different algorithms for solving the same problem. (c) A formula. (d) Used broadly in methodological writing to mean any step-by-step procedure to solve a problem. Named after the 9th-century Arab mathematician and inventor of algebra, Al-Khwarizmi, whose name in Latin was Algorithmi.

Alienation, Coefficient of   (a) A measure of the lack of relationship between two *variables. It is sometimes symbolized as k or k^2; usage, and therefore the meaning, varies. When the coefficient of alienation is high, it is hard to make predictions about one variable by knowing the value of another. The stronger the correlation between two variables, the weaker the coefficient of alienation will be. (b) A measure of how well a *model fits actual data. The lower the coefficient, the better the fit.

The coefficient of alienation is equal to 1 minus the coefficient of *determination, that is, to $1 - r^2$ or $1 - R^2$ or to 1 minus the square root of the coefficient of determination, that is, $1 - r$ or $1 - R$. Compare coefficient of *nondetermination.

Allocation Rule   In *discriminant analysis, a rule for deciding in which category to place a new observation.

A

Alpha [A, α]   (a) Usually called *Cronbach's alpha to distinguish it from the alpha in *alpha level. It is a measure of internal *reliability of the items in an *index. Cronbach's alpha ranges from 0 to 1.0 and indicates how much the items in an index are intercorrelated and thus presumed to measure the same thing. (b) Symbol for the *intercept in a *regression equation. (c) Symbol for an *odds ratio. (d) Symbol for "is proportional to."

Alpha Error   An error made by rejecting a true *null hypothesis (such as claiming that a relationship exists when it does not). Also called *Type I Error. See *beta error, *hypothesis testing.

Alpha Level   (a) The chance a researcher is willing to take of committing an *alpha error or *Type I error, that is, of rejecting a *null hypothesis that is true. (b) The probability that a Type I error (wrongly rejecting the null hypothesis) has been committed.

   The smaller the alpha level, the more significant the finding, because the smaller the likelihood that the finding is due to chance (*random error). Thus, an alpha level of .01 is a more difficult criterion to satisfy than a level of .05. Also called *level of (statistical) significance. See *p value, *probability value, *probability level, *sampling error.

Alphanumeric Variable   A variable that can be expressed as a letter, number, other symbol, or some combination of letters, numbers, and symbols.

Alternative Hypothesis   In *hypothesis testing, any hypothesis that does not conform to the one being tested, usually the opposite of the *null hypothesis. Also called the *research hypothesis. Rejecting the null hypothesis shows that the alternative (or research) hypothesis may be true. Symbolized H_1 or H_a.

   For example, researchers conducting a study of the relation between teenage drug use and teenage suicide would probably use a null hypothesis something like: "There is no difference between the suicide rates of teenagers who use drugs and those who do not." The alternative hypothesis might be: "Drug use by teenagers increases their likelihood of committing suicide." Finding evidence that allowed the rejection of the null hypothesis (that there was no difference) would increase the researchers' confidence in the probability that the alternative hypothesis was true.

Alternative Methodologies   A catchall term referring to any *nonexperimental and/or nonquantitative research methods. See *qualitative research.

AMOS   Analysis of Moment Structures. A *statistical package used especially for *structural equation models.

Anachronism   An error made by affirming something that is chronologically impossible, such as by attributing *cause to something that followed the effect. Compare *ahistorical.

For example, to claim that Keynes's writings influenced Adam Smith's theories would be an anachronistic mistake; it is impossible because Smith died long before Keynes wrote anything.

Analog (Also spelled: analogue) Said of data or computers that use a system of representation that is physically analogous to the thing being represented. Compare *digital.

For example, a thermometer can use the height of a column of mercury to indicate heat; the higher the column, the higher the temperature. Or, analog watches use the physical movement of hour and minute hands to represent the passing of time; the more the hands have moved, the more time has passed.

Analysis (a) The separation of a whole into its parts so as to study them. (b) The study of the elements of a whole and their relationships. (c) Loosely, but perhaps most commonly, any rigorous study of anything. (d) In the context of *design and *measurement, the term is used broadly to mean statistical techniques. Compare *synthesis.

Analysis of Covariance (ANCOVA) An extension of *ANOVA that provides a way of statistically *controlling the (*linear) effects of variables one does not want to examine in a study. These *extraneous variables are called *covariates, or control variables. ANCOVA allows you to remove covariates from the list of possible explanations of variance in the *dependent variable. ANCOVA does this by using statistical techniques (such as *regression) to *partial out the effects of covariates rather than direct experimental methods to control extraneous variables.

ANCOVA is used in experimental studies when researchers want to remove the effects of some *antecedent variable. For example, pretest scores are used as covariates in pre-/posttest experimental designs. ANCOVA is also used in *nonexperimental research, such as surveys of nonrandom samples, or in *quasi-experiments when subjects cannot be assigned randomly to *control and *experimental groups. Although fairly widespread, the use of ANCOVA for *nonexperimental research is controversial. All ANCOVA problems can be handled with *multiple regression analysis using *dummy coding for the nominal variables, and, with the advent of powerful computers, this is a more efficient approach. Because of this, ANCOVA is now used less frequently than in the past.

Analysis of Covariance Structures An alternate term for *structural equation modeling (SEM).

Analysis of Variance (ANOVA) A test of the *statistical significance of the differences among the *mean scores of two or more groups on one or more *variables or *factors. It is an extension of the *t-test, which can handle only two groups at a time, to a larger number of groups. More specifically, it is used for assessing the statistical significance of the relationship between *categorical *independent variables and a *continuous *dependent variable. The procedure in ANOVA involves computing a ratio (*F ratio) of the *variance between the groups (*explained variance) to the variance within the groups (*error variance). See *one-way ANOVA, *two-way ANOVA. ANOVA is equivalent to *multiple regression with *dummy-coded *independent variables and a continuous dependent variable.

For example, a professor tried different teaching methods. He randomly assigned members of his class of 30 students to three groups of 10 students each. All three groups were given the same required readings, but class time was spent differently in each. Group 1 ("Discuss") spent class time in directed discussions of the assigned readings. Group 2 ("No Class") was excused from any obligations to attend classes for the first half of the semester, but they were given additional text materials they could use to help them understand the assigned readings. Group 3 ("Lecture") was taught by traditional lecture methods. The students' scores on the midterm examination are listed in Table A.1. Students in the three groups obviously got different average scores. The professor wanted to know whether the differences were statistically significant, that is, whether they were bigger than would be likely due to chance alone. To find out, he entered the information from Table A.1 into his computer and conducted an ANOVA. As Table A.2

Table A.1   Analysis of Variance: Students' Midterm Scores, by Method of Instruction

	Group 1 Discuss	Group 2 No Class	Group 3 Lecture
	94	78	87
	92	76	85
	91	72	84
	89	71	84
	88	68	81
	88	68	80
	86	67	80
	86	66	79
	83	64	72
	83	60	68
Total	880	690	800
Mean	88	69	80

Table A.2   ANOVA Summary Table: Three Teaching Methods

Source	SS	Df	MS	F
Between Groups	1820	2	910.00	35.10*
Within Groups	700	27	25.93	
Total	2520	29		

* p < .001.

shows, the results were highly statistically significant (at the p < .001 level); the teaching methods almost certainly made a difference. Compare *factorial experiment, *two-way ANOVA.

How to Read an ANOVA Summary Table. "Source" means source of the variance. "Between groups" is explained variance, that is, explained by the treatments the different groups received. "Within groups" is unexplained or error variance, since differences among individuals within a group cannot be explained by differences in the treatments the groups received. "SS" is *sum of squares (total of squared *deviation scores). *Degrees of freedom are abbreviated to "df." "MS" stands for mean squares, which are calculated by dividing the SS by the df. "F" is the F ratio of the MS between to the MS within, which is statistically significant at the .001 level (p < .001). (Note that it is common for such tables to be cut from research reports in the name of saving space.)

Analytic Induction   An approach in *qualitative research that develops theory by examining a small number of cases. Theory then leads to formulation of a hypothesis, which is tested through the study of more cases. This usually leads to refinement or reformulation of the hypothesis, which is then tested with further cases until the researcher judges that the inquiry can be concluded. Compare *negative case analysis.

ANCOVA   *Analysis of covariance.

Anecdotal Evidence   Evidence, often in story form, derived from casual, unsystematic, and/or uncontrolled observation. Usually used to dismiss someone's evidence, as in *"mere* anecdotal evidence." Compare *case study, *narrative analysis.

Anonymity   The assurance that individual participants in research cannot be identified. The strictest kind of anonymity means that the identity of subjects is concealed even from researchers, as is possible when respondents return surveys by mail. Many kinds of research, such as face-to-face interviews, make strict anonymity impossible. Then *confidentiality* of information about individual participants, including their identities, becomes the

A

standard. While anonymity means that the researcher does not know subjects' identities, confidentiality means that the researcher knows but promises not to tell. See *research ethics, *institutional review board.

ANOVA   *Analysis of variance.

Antecedent   (a) A condition that precedes another and is thought to influence or to cause it. (b) The first term in a ratio. For example, in the ratio 14:1, 14 is the antecedent; 1 is the *consequent.

Antecedent Variable   A variable that comes earlier in an explanation, or in a chain of causal links—as in a *path analysis.

   For example, if the dependent variable were occupation at age 50, education level would usually be an antecedent variable, and place of birth would be a variable antecedent to education.

Anthropometry   Research that involves measurements of the human body.

Antilog   See *logarithm.

Antimode   The least common score or value in a distribution. Compare *mode.

Antinaturalism   The belief that using methods developed in the natural sciences is inappropriate in the study of human thought and action. Contrast *positivism.

A Posteriori   Literally, Latin for "from what comes after"; said of conclusions reached by reasoning from observed facts (after observation) or of research that proceeds in an *inductive way. Loosely, *empirical. Compare *a priori, *post hoc comparisons.

A Posteriori Comparison   A comparison that a researcher decides to make after the data have been collected and studied. This is usually done because the results have suggested a new way to approach the data. See *post hoc comparison, *multiple comparisons. Compare *a priori comparison.

Applications Software   Computer programs designed for specific purposes (applications) such as word processing, accounting, or doing statistics problems.

Applied Research   Research undertaken with the intention of applying the results to some specific problem, such as studying the effects of different methods of law enforcement on crime rates. One of the biggest differences between applied and *basic research is that in applied work the research questions are more often determined not by researchers, but by policymakers or others who want help. Virtually any *research design can be, and has been, used to conduct applied research. Types of applied research include *evaluation research and *action research.

A

Approximation Error   Another term for *rounding error.

A Priori   Literally, Latin for "from what comes before." (a) Used to describe preexisting (prior) conditions among groups of subjects, especially potential *confounding variables. (b) Said of conclusions reached on the basis of reasoning from self-evident propositions—without or before examining facts—or of research that proceeds in a *deductive way. Loosely, theoretical. Compare *a posteriori.

The expression is often used to describe a conclusion for which someone believes there is no empirical evidence, as in: "One conclusion is as likely to be true as the other, a priori, which is why we need to gather more data to resolve the question."

A Priori Comparison   A comparison that a researcher decides to make before (prior to) performing the experiment or gathering the data. Designs incorporating a priori comparisons are usually considered stronger than those that use *a posteriori comparisons. See *planned comparison, *Bonferroni test statistic.

A Priori Probability   Another term for *theoretical probability. Contrast *prior probability.

Aptitude-Treatment Interaction   A *trait-treatment interaction research design in which the trait is an aptitude, such as verbal ability or manual dexterity. Aptitude-treatment interaction was the original term; *attribute-treatment and *trait-treatment were more general versions developed later.

Archive   (often used in the plural) (a) A place where public records or other kinds of information are stored. (b) The information thus stored.

*Database archives (such as the *Public Use Microdata Samples and the *General Social Survey) are becoming increasingly important in social and behavioral research. Qualitative data archives have a very long history. Recent technical advances (e.g., videotapes of interviews) have increased the scope and public availability of such archives.

Archival Research   The study of existing records (archives). Often associated with historical research because, by definition, existing records will contain information about the past.

Arc-Sine Transformation   A statistical transformation used to increase the homogeneity of variance of samples, that is, to make sample variances more similar across comparison groups. This transformation is done because equality of variances is a fundamental *assumption of several important statistical tests, such as *ANOVA. The transformation is used to change data made up of frequencies or proportions, so that they can be more accurately analyzed with ANOVA or regression.

A

Area Sample   A kind of *cluster sample in which the clusters are selected on the basis of geographic units. Area samples use maps rather than lists to define the *sampling frames from which the sample is then drawn.

For example, say you want to survey residents of California about their *attitudes concerning property taxes. You do not have a list of all California residents from which to draw a sample. You could divide the state into "areas" (such as *census tracts or voting districts) and take a *random sample of those areas. Then, within each of these selected areas, you could survey all (or a sample) of the residents on their attitudes.

ARIMA   Autoregressive Integrated Moving Average. A complex set of statistical techniques for *time-series analysis, which combines *autoregressive analysis procedures with those of *moving averages. By using ARIMA models, the researcher is able to construct *trend lines that take into account both the *systematic error (autocorrelation) and the unsystematic or *random error (or *noise).

Well-known examples of findings based on this sort of work are the *seasonally adjusted figures for unemployment or consumer spending.

Arithmetic Mean   See *mean.

Array   (a) An ordered display of a set of observations, measurements, or statistics—such as a *frequency distribution. For example, the observed grade-point averages 3.2, 2.2, 3.1, 3.6, 2.8 in *ascending order give the array {2.2, 2.8, 3.1, 3.2, 3.6}. (b) An arrangement of numbers in a *matrix. See *correlation matrix.

Arrow's Impossibility Theorem   A logical demonstration by the economist Kenneth Arrow that, when there are three or more alternatives, there is no way of deciding by voting how to consistently combine the distinct preferences of individuals in such a way that each of their preferences will be maximized. In short, it is impossible for society to make up its mind democratically about what it wants. The theorem is generally accepted as true and important in economics, social theory, political science, and *decision theory.

Artifact   An artificial result. (a) A mistaken or *biased result produced by the measuring instrument rather than by the phenomenon being studied. (b) Something the researcher created by the way he or she gathered or analyzed the *data, not by a condition present in the *subjects. Examples include *halo effect, *Hawthorne Effect, *John Henry Effect, *regression artifact, and *demand characteristics. See *self-fulfilling prophecy.

Artificial Intelligence   A branch of computer science that studies ways of getting computers to simulate human thought, including such abilities as reasoning and learning from experience.

Ascending Order   Said of data arranged so that each item in a series is higher than (ascends) the previous items. Arranging data in this way can be an important step toward constructing a *frequency distribution or calculating a *rank order correlation. Compare *descending order.

ASCII   (rhymes with "passkey") American Standard Code for Information Interchange. This computer character set is used as a sort of Esperanto for *software. It enables a user of one software *program to translate what he or she has written into a language that can be understood by another software program. ASCII is widely used to transfer statistical data files from one computer or program to another.

Assessment Research   Often a synonym for *evaluation research. When a distinction is drawn between the two, assessment frequently refers to measuring individual outcomes, while evaluation refers to studying the effects of programs. The two are routinely linked, because a common way to evaluate a program is to assess its effects on the individuals who participated in it.

Association, Measure of   Any *statistic that shows (in a single number) the degree of relationship between two or more *variables. There are two broad types: those based on *proportional reduction of error in prediction, such as *lambda, and those based on departure from *statistical independence, such as *phi.

   Other examples of measures of association include *Pearson's correlation coefficient, *gamma, and *Cramer's *V.* Note that there can be considerable controversy about which measures of association are best to use for various purposes and kinds of data.

Association, Statistical   (a) A relationship between two or more *variables that can be described statistically. (b) Any of several statistical techniques (such as *correlations and *regression analysis) that can be used to describe the degree to which differences in one variable are accompanied by (associated with) corresponding differences in another variable. Such associations do not prove the presence of a causal relation, but they suggest that one may be present.

Association, Test of   Another term for *test statistic when applied to a *measure* of association, which indicates the size of the relation between two variables. Like any test statistic, a test of association tests for *statistical significance, that is, it estimates how likely it is that an association of a given size, in a sample of a given size, would have occurred had there been no association in the population from which the sample was drawn.

Association Models   In the analysis of *categorical data, models that analyze the observed frequencies in cross-classified tables so as to measure the strength of association between two or more ordered categorical variables. Compare *log-linear models, of which association models are a subcategory.

Assumption (a) A statement that is presumed to be true, often only temporarily or for a specific purpose, such as building a *theory. Compare *axiom, *hypothesis. (b) The conditions under which statistical techniques yield valid results.

For example, (a) researchers might make the assumption that there is no difference in the innate verbal ability of men and women so that they could use differences in grades in college-level literature courses to test a theory that college professors discriminate on the basis of sex.

For example, (b) most statistical techniques, such as *regression analysis and *ANOVA, require that certain assumptions (e.g., *homoscedasticity) be made about the data. Serious violations of the assumptions can make the results misleading or meaningless. Tests of *statistical significance assume random samples or *random assignment to control and experimental groups, and are hard to interpret if those assumptions are not met.

Asymmetrical Distribution A *skewed distribution.

Asymmetrical Test Another term for *one-sided test of significance.

Asymmetric Measure A measure of *association that has a different value depending on which *variable is treated as *dependent and which as *independent. Compare *symmetric measure.

*Lambda and *Somers's d are examples of asymmetric measures of association.

Asymmetry The state of a *distribution that is *skewed or unbalanced.

Asymptote (a) A theoretical limit that a curve approaches but never reaches. (b) Any theoretical outer limit; see *limit. (c) A leveling off of performance quality or efficiency; see *learning curve.

Asymptotic Said of a curve that gets ever closer to a line but never touches it. For example, the *tails of a *normal curve are asymptotic to the x axis.

Attenuation A reduction in a measure of *association caused by measurement errors. Compare *artifact. Since no measurement is without error, attenuation is all but universal.

For example, in a study of consumer confidence and spending, neither measure will be perfect, which means that a correlation between confidence and spending will understate the true strength of the relation.

Attenuation, Correction for An adjustment made to *correlation coefficients to estimate more accurately what they would have been had errorless measurements been available. The greater the *reliability of the measures, the smaller the attenuation. The lower the reliability, the greater the attenuation and the correction.

A

Attitude   A positive or negative evaluation of and disposition toward persons, groups, policies, or other objects of attention. Attitudes are learned and relatively persistent. See *trait, *Bogardus Social Distance Scale.

Attitude Scale   A series of questions designed to measure the strength of attitudes and beliefs. A common format for such measures is a *Likert scale. Compare *Guttman scale, *index.

Attribute   (a) A *qualitative variable or trait, often used in contrast with a *quantitative variable or trait. This is the same basic distinction as that between *categorical (attribute) and *continuous (quantitative) variables. Compare *attribute variable.

For example, a person's sex is an attribute (or qualitative, categorical variable), while the number of square feet in his or her living room is a quantitative, continuous variable. Occupation is a variable; carpenter, teacher, and butcher are attributes.

(b) In *conditional probability, the "given" or condition. For example, for the problem, What is the probability that a college sophomore will graduate given that she is on the dean's list? being on the dean's list is the attribute.

Attribute-Treatment Interaction (ATI)   Another term for *trait-treatment interaction and *aptitude-treatment interaction. The purpose of studying an ATI is to learn whether the effect of an attribute on a *dependent variable is the same or different for two or more groups that have received different *treatments. See *interaction effect.

Attribute Variable   A variable that is a characteristic or trait of a subject which, therefore, researchers cannot manipulate but can only measure.

For example, in a study of the effects of vitamins on health, researchers could vary (manipulate) the amount and kind of vitamins subjects received, but they could only measure, and not manipulate, the subjects' ages.

Attrition   Losing subjects over the course of the research project. Also called "mortality." Attrition is a common problem in *longitudinal studies and may be a source of *bias if the subjects who are lost make the *sample less *representative of the *population.

Attrition is a frequent issue in *panel studies, in which the same subjects are studied at two or more times. Attrition may occur, for instance, when subjects move and cannot be located. If many subjects are lost, and if they are in some way unusual in comparison to the remaining subjects, this makes it hard to draw valid conclusions about what happened to the group over time.

Audit   Inspection of an organization's or an individual's records. The term was originally used to describe reviewing financial records, but it is increasingly used in social and behavioral research, especially in qualitative research, to mean examining the extent to which the methods of a study are sound and its findings confirmable. Audits can be external (done by an outside expert) or internal as a self-check on efficiency and effectiveness.

Audit Trail   A systematic set of records that allows someone conducting an *audit to reconstruct what was done, such as keeping track of changes made in a data file. The term is mainly used to describe documentation that allows financial transactions to be traced, but it is also used to describe researchers' records, especially in qualitative research. Sometimes called a "paper trail." Compare *protocol.

Authenticity Criteria   Standards appropriate for judging the quality of research. Used especially in the context of *qualitative, *constructivist, and *postmodernist research.

Autochthonous Variability   Change that comes from influences within a causal system as opposed to from outside it. Compare *endogenous and *exogenous variables.

Autocorrelation   Correlation between members of a series of observations, such as weekly oil prices or interest rates. Autocorrelation is present when later items in a *time series are correlated with earlier items. Autocorrelations are zero when the time series is random. More technically, autocorrelation occurs when *residual *error terms from observations of the same variable at different times are correlated. Such correlations can raise several kinds of interpretive problems. In *regression analysis of time-series data, for example, autocorrelation may arise from the tendency of effects to persist over time, even when *independent variables change. In regression analysis, autocorrelation can be reduced by using *generalized rather than *ordinary least squares to compute the *regression equation. Also called "serial correlation." See *multicollinearity, *ARIMA.

For example, declining interest rates usually lead people to buy more on credit; but lower interest rates do not always produce the expected effect, at least in the short run, and certainly not immediately. People's credit purchases (dependent variable) may continue unchanged, even when economic conditions (independent variables) that "ought" to increase them have changed. If there are autocorrelations between credit purchases at week 1 and week 2, and week 2 and week 3, and so on, this will obscure any long-term relation between interest rates and credit purchases.

Autoregressive   Said of a series of observations in which the value of each depends (at least in part) on the value of one or more of the immediately preceding observations. Called autoregressive because one explains later observations by earlier ones, that is, one *regresses later values on earlier values. See *autocorrelation, *Markov chain.

A

Average   See *mean.

Average Deviation (AD)   A measure of the *variation in a group of scores. It is calculated by taking the *mean or average of the *absolute values of the *deviation scores (that is, the differences between the scores and their mean). The larger the average deviation, the greater the *spread of scores in a group of scores. An AD can also be, and often is, calculated on the basis of deviations from the *median. The AD is less commonly used as a measure of variation than are the *variance and the *standard deviation. Also called "mean deviation" and, most accurately, *mean absolute deviation.

Axiom   A maxim or statement that is considered so accurate or self-evident that it is widely accepted as a foundation on which arguments can be built, or a truth from which other truths can be deduced. Compare *assumption, *postulate.

In contrast to mathematics, there are few genuine axioms in the social and behavioral sciences. Two statements that might qualify as axioms (at least for some researchers) are: (1) out-group hostility breeds in-group solidarity; (2) people seek to maximize pleasure and minimize pain.

Axis   A vertical or a horizontal line used to construct a *graph. See *abscissa (*x* axis), *ordinate (*y* axis), *Cartesian coordinates.

B   Symbol for an unstandardized *regression coefficient; usually a lowercase, italicized *b*.

Background Variables   Aspects of subjects' "backgrounds" that may influence other variables, but will not be influenced *by* them. Background variables are usually demographic characteristics—such as age, sex, ethnicity, and parents' income—that the researcher cannot manipulate. Also called "subject," "organismic," "classification," and "individual-difference" variables. While there can be subtle differences among these usages, the basic idea is the same: Background variables can be causes, but they are not *independent variables, in some strict senses of that term, because they cannot be manipulated by a researcher; however, researchers can and frequently do *control for background variables. Compare *antecedent variable.

For example, your sex and your age (background variables) might influence your income (think of age and sex discrimination), but a change in your income certainly would not change either your sex or your age.

Backward Elimination   A computer procedure for *regression analysis that is used to identify the *independent variables that are good predictors of the *dependent variable in order to find the best *fitting equation or model. Also called "stepdown selection." Compare *forward selection and *stepwise regression.

The routine is to begin the analysis with all the variables in the equation and remove (eliminate) them one at a time according to whether they meet specific criteria (levels of significance of their *F ratios). The variable with the smallest *partial correlation is examined first. If it does not meet the criterion, it is eliminated; then the variable with the second smallest partial is examined, and so on until no more variables are eliminated.

B

Balanced Design   A *randomized-blocks design in which every *treatment appears in each block the same number of times. The same number of observations is made for each experimental condition.

Bar Chart   Another term for a *bar graph. Also called "bar diagram."

Bar Graph   A way of depicting *frequency distributions for *categorical (*nominal) variables, such as religious affiliation, ethnic group, or state of residence.

   The following example presents the Republican Party affiliation of a sample of survey respondents categorized by religious affiliation; about 31% of Protestants surveyed reported affiliation with the Republicans, roughly 17% of Catholics did, and so on. Note that the bars do not touch in a bar graph as they do in a *histogram.

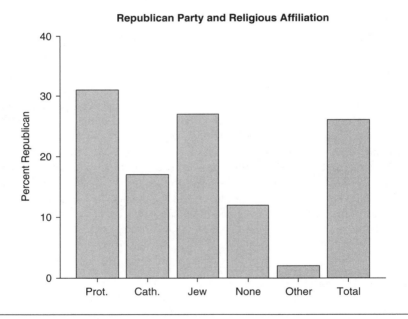

Figure B.1   Bar Graph

Bartlett-Box Test   One of several tests for *equality of variance. *Cochran's *C* test is another. Also called Bartlett-Box *F* test.

Barlett's Test   A test of the *sphericity assumption in a *factor analysis. It tests the *null hypothesis that all the correlations in the matrix to be factor-analyzed are zero. If they are not significantly different from zero, a factor analysis is inappropriate. See *KMO test.

B

Baseline   (a) The average rate or level of some *variable before an experimental *treatment is applied or before some measure of the effect of a new variable is taken. Compare *pretest. (b) The horizontal, or *x axis, that is, the *abscissa.

For example, (definition a) if a political party wished to test the effects on party finances of a new advertising campaign, it could get baseline data by measuring the rate of contributions a week before the campaign started ("the baseline") and comparing that with the rate of contributions for the week after the campaign began.

Basement Effect   Another term for *floor effect.

Base Year (or Date or Period)   The first in a series of years (or periods or dates) in a *time-series index. Often the base-year value is set at 100; subsequent changes can be easily seen as a percentage of the base year.

For example, if a price index were set at 100 for the base year of 1998, and the index moved to 140 by 2005, this means that prices went up by 40%.

BASIC   Beginner's All-Purpose Symbolic Instruction Code. A *programming language, intended for nonspecialists, widely used for writing microcomputer programs.

Basic Research   Research undertaken with the primary goal of advancing knowledge and theoretical understanding rather than solving practical problems. Often contrasted with *applied research. Sometimes called *pure research.

Baud   A measure of how quickly digital data can be transmitted; often the speed at which a *modem can send *bits over a telephone line. For example, 1,200 baud equals 150 characters (*bytes) per second, 2,400 equals 300, and so on.

Bathtub Curve   In studies of *life expectancy, the death rate often looks roughly like a cross section of a bathtub. The death rate starts fairly high, declines rapidly until about age 5, then levels off for some decades before starting to climb rapidly again around age 50. See *life table.

Battery of Tests   A group or series of tests, usually psychological tests.

Bayesian Decision Theory   A method of decision making, and by extension, of estimating probabilities, that is based at least partly on expert opinion, called "subjective" probability.

Bayesian Inference   Statistical inference based on *Bayes's theorem and on the researcher's *subjective beliefs about the topic being studied. Named after Thomas Bayes, an 18th-century English mathematician. His method of inference is controversial, although his theorem is not. The method of inference involves working "backwards," from effect to *cause, by estimating the *conditional probability of a cause—given that certain events (effects) have occurred. In other words, Bayesian inference involves using new information, called the *posterior probability, to revise earlier estimates of probability, called *prior probability. The "Empirical Bayes Method" uses data, rather than subjective opinion, to estimate the prior probability.

Bayesian Linear Models   A term occasionally used to describe *hierarchical linear models.

Bayes's Theorem   A method for evaluating the *conditional probability of an *event, particularly an unknown prior event, given that a known subsequent event has occurred. Bayes's theorem allows researchers to revise estimates of the probability of events using new evidence to do so. See *Bayesian inference.

Before-After Design   Any research design in which the subjects are given a *pretest and a *posttest. See *A-B-A-B designs, *repeated-measures designs.

Behavioral Sciences   Disciplines that study the actions (behaviors) of human beings and other animals. Commonly included in lists of the behavioral sciences are psychology, sociology, social anthropology, economics, and political science.

Behavior Coding   Methods assessing the quality of *surveys by observing and coding the behaviors of interviewers and *respondents, such as whether the interviewer reads the question accurately and whether the respondent asks for clarification or refuses to answer.

Behaviorism   (a) A theoretical position in psychology and related disciplines contending that the only scientific subject matter is behavior—not beliefs, attitudes, desires, or other mental states. It is most commonly associated with the work of John B. Watson in the early 20th century and B. F. Skinner more recently. (b) Advocacy of research based on empirical, external observation of behaviors.

Behrens-Fisher Test   An extension of the *t-test of the *statistical significance of the difference between two means; it relaxes the requirement of equal population *variances. It is somewhat controversial and not widely used.

Bell-Shaped Curve   A symmetrical curve, usually plotting a continuous *frequency distribution, such as a *normal distribution, which looks like a cross section of a bell. See *normal distribution for an illustration. The *Student's t distribution is also bell-shaped, although it is not often referred to that way.

Bernoulli Distribution   Another name for the *binomial distribution.

Bernoulli Hypothesis   The hypothesis that a person's decision about whether or not to take a risk is based not just on the probability of success but also on the value attached to the thing being risked (usually called *utility).

For example, if a person were considering investing $10,000, her decision would be based on the probability of losing the money and of making a profit. But it would also likely be based on how valuable or important the money was to her. If the $10,000 was her life's savings and all that kept her from starvation, her willingness to take a small risk (even with a strong probability of a large return on the investment) might not be very great.

Bernoulli Process   Two classes of events and their associated probabilities. See *Bernoulli trial.

Bernoulli's Theorem   In a *probability experiment, the larger the number of trials, the closer the *empirical probability will come to the *theoretical probability.

For example, the more flips of a fair coin, the closer the actual percentage of heads will be to the theoretical probability of 50%.

Bernoulli Trial   In *probability theory, a trial or experiment with two possible outcomes, such as heads/tails, win/lose, 7/not 7. One of the two is usually called "success," and its probability is termed "p." The other is called "failure," and its probability is labeled "q." Since $p + q = 1.00$; $1 - p = q$; $1 - q = p$. Named after the Swiss mathematician, Jacques Bernoulli (1654–1705).

Best Fit   See *goodness-of-fit.

Best Estimator   Any method of estimating a *population *parameter using specific optimizing techniques for calculations on the sample, such as the *least-squares criterion or *maximum likelihood methods. See *best linear unbiased estimator.

Best Linear Unbiased Estimator (BLUE)   A *regression line computed using the *least-squares criterion when none of the *assumptions is violated. Abbreviated BLUE. A BLUE will have a smaller variance than any other estimator of the *population parameter.

Beta [B, β]   Greek letter used to symbolize several statistical concepts, including (a) *Type II errors, (b) standardized *regression coefficients, and (c) *population parameters of unstandardized regression coefficients.

Beta Coefficient   (a) A *regression coefficient for a *sample expressed in *standard deviation units (i.e., *z-scores). Specifically, the beta coefficient

B

indicates the difference in a *dependent variable associated with an increase (or decrease) of one standard deviation in an *independent variable—when *controlling for the effects of other independent variables. Also called *standardized regression coefficient and *beta weight. (b) A statistic summarizing the movement in the price of a particular stock compared to that of the stock market as a whole. The bigger the beta coefficient, the greater the volatility of the stock.

Note: A *regression coefficient expressed in nonstandardized units is usually symbolized by b. Usage is confusing, because beta is also used to symbolize the population parameter of b.

Beta Error   An error made by accepting or retaining a false *null hypothesis—more precisely, by failing to reject a false null hypothesis. This might involve, for example, claiming that a relationship does not exist when in fact it does. Also called *Type II Error. Compare *alpha error. See *hypothesis testing.

Beta Level   The probability of making a *beta error, that is, failing to reject a false *null hypothesis. Compare *alpha level.

Beta Weights   Another term for *standardized regression coefficients, or *beta coefficients. Beta weights enable researchers to compare the size of the influence of *independent variables measured using different *metrics or scales of measurement. Also called "regression weights."

For example, imagine a *regression analysis studying the influence of age and income on attitudes. Subjects could be adults ranging in age from 18 to 80. Their incomes might vary from $4,000 for a high school senior working after school to $200,000 for a tax lawyer. By reporting years and dollars as *standard scores, rather than in the original metric, beta weights allow the researcher to make easier comparisons of the relative influence of age and income on attitudes.

Between-Group Differences   Usually contrasted with differences within the groups being studied in an *analysis of variance (ANOVA). Between-group differences are what the researcher is interested in; they are considered large only if they are large in comparison to *within-group differences. Also called "interclass variance."

For example, the results of an experiment are reported in the following table. The between-group difference (between the means of the groups) appears significant since it seems quite a lot larger than the differences among subjects within each group. Using ANOVA or a t-test to test for statistical significance indicates that the means in Table B.1 are significantly different ($p < .001$).

Table B.1   Scores Illustrating Between-Group Differences

	Control Group	Experimental Group
	72	91
	71	89
	68	88
	67	86
	66	85
	64	83
Total	408	522
Mean	68	87

Between-Subjects Design (or ANOVA)   A research procedure that compares different subjects. Each score in the study comes from a different subject. Usually contrasted to a *within-subjects design, which compares the same subjects at different times or under different *treatments.

Between-Subjects Variable (or Factor)   An *independent variable or *factor for which each subject is measured at only one *level or under one *condition. See *within-subjects variable, *repeated-measures designs, *nested design.

Between Sum of Squares   A measure of *between-group differences. It is calculated by squaring and summing *deviation scores. It is used in comparison to *within-group differences to compute the *F ratio in an *analysis of variance. Symbolized $SS_{between}$. See *mean squares.

Bias   (a) Anything that produces *systematic* error in a research finding. By contrast, *random* error tends to balance out in the long run; the distortions due to systematic error continue to grow in the long run. See *biased estimator. (b) Also, the effects of any factor that the researcher did not expect to influence the *dependent variable.

   For example, suppose you wanted to survey the opinions of New York City residents (the population). If you stood on a busy street corner at noon and asked the first 200 people who walked by to respond to your survey, your results would almost surely be biased. Perhaps your corner is near a hotel where out-of-town conventioneers usually stay, or your corner might be one where only very poor people are found, or where men seldom pass by at that hour. You would be very lucky (and you would never really know) if some such factor were at work to bias your results by making your 200 respondents unrepresentative of the general population of the city.

Biased Estimator   When the *expected value of a *sample statistic tends to over- or underestimate a *population parameter, it is called a biased estimator.

For example, the *standard deviation (SD) of a sample is a biased estimator; it underestimates the population standard deviation. To correct for that bias, when computing the SD, the *sum of squares is divided by $n - 1$, rather than n. This correction is not necessary when computing the SD of a population; it is needed only when using a sample statistic to estimate a population parameter.

Biased Sample   A sample selected by a nonrandom process. Compare *probability sample.

Bifactor Model   In *factor analysis, the assumption that the *variance can be analyzed into a general factor and some *mutually exclusive subfactors.

Bimodal Distribution   A distribution having two *modes or peaks. Strictly speaking, to call a distribution bimodal, the peaks should be the same height. However, it is quite common to call any two-humped distribution bimodal, even when the high points are not exactly equal.

In the following illustration, the number of students getting various scores on a 12-item test is plotted on the graph. Nineteen students got a score of 4 correct; another 19 got 8 right.

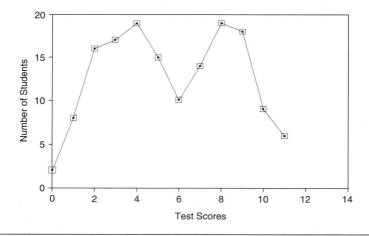

Figure B.2   Bimodal Distribution

Binary   (a) An adjective describing a number system or coding system that uses only two digits, generally 0 and 1. (b) A binary variable is one that can have only two possible values, such as sex: male or female; these are usually coded 0 and 1.

Binomial Distribution   A *probability distribution for a *dichotomous or two-value *variable (binomial = "two names"), such as success/failure, profit/loss, or in/out. Also called "Bernoulli distribution."

Table B.2   Binomial Distribution: Probability Distribution for Drawing Clubs

B

Number of Clubs	Probability
0	.2373
1	.3955
2	.2637
3	.0879
4	.0146
5	.0010

For example, suppose you take an ordinary deck of playing cards, shuffle it, draw a card, note what you draw, replace the card, and repeat the process five times. Say you record only one of two events: getting a club or not (clubs/not clubs). The probability distribution is given in Table B.2. Reading it, you can see that the probability of getting clubs on all five draws is very small (.0010). The probability of getting exactly one club is .3955. To calculate the probability of getting two or more clubs, add together the probabilities of getting exactly 2, 3, 4, and 5, which equals .3672.

Binomial Probability   (a) A kind of probability calculation used when there are only two possible outcomes (binomial = "two names") for each of a series of trials, such as heads/tails, win/lose, true/false. (b) The chances associated with a series of trials when there are only two possible outcomes. See *binomial distribution.

For example, suppose you are taking a true/false test tomorrow. There will be a total of 25 questions. To pass you need to get at least 18 correct. You could use binomial probability calculations to figure out how likely you are to pass by *random guessing alone.

Binomial Test   A test of *statistical significance of an outcome that proceeds by calculating how likely (*p value) a particular result is, that is, how much it differs from chance as described by the *binomial distribution.

Binomial Variable   A variable with only two values or two "names" (a *dichotomous variable), such as strong/weak, left/right, rich/poor, male/female.

Biometrika   Pioneering British statistics journal founded in 1901 by Karl Pearson (of the *Pearson's correlation) in which many of the statistical techniques most widely used today were first described. Many statistical tables (e.g., *F*, *t*, and chi-squared distributions) reprinted in the backs of textbooks were originally put together by the editors of that journal.

Biostatistics   Statistics applied to biological research, often research in medicine and *epidemiology.

B

Biplot   A graphic technique for depicting both the values of variables and the differences among individuals.

Bipolar Factor   In *factor analysis, a factor that has both positive and negative *factor loadings. Such factors are usually difficult to interpret.

Bipolar Scale   A scale in which the opposite ends represent contrasting concepts, such as lazy and industrious. A *semantic differential scale is bipolar.

Biquartimin   A method of (*oblique) rotation of the *axes in a *factor analysis.

Biserial Correlation   A *correlation coefficient computed between a *dichotomous and a *continuous variable. The dichotomous variable is actually an interval-level variable, but one that has been *collapsed to only two levels (such as high and low). The biserial correlation provides an estimate of what the correlation would have been if the collapsed dichotomous variable had been left as a continuous variable. The estimate is usually high. Abbreviated r_{bis}. Compare *tetrachoric correlation, *point biserial correlation.

Biserial *r*   See *biserial correlation.

Bit   A binary digit (1 or 0). It is the smallest unit of information recognized by a computer. Several bits (usually 8) are combined to make a *byte.

Bivariate   Pertaining to two *variables only.

Bivariate Association   A relation (*covariation) between two variables only. Among the many measures of bivariate association are *eta, *gamma, *lambda, *Pearson's *r*, *Kendall's tau, and *Spearman's rho.

Bivariate Distribution   The joint distribution of two variables. In a bivariate distribution each member of a particular sample or population has a score on two variables—say, height (X) and weight (Y). This makes for three distributions: the distribution of height, the distribution of weight, and the distribution of the *covariance of height and weight, which is the bivariate distribution.

When a bivariate distribution is discussed, the bivariate normal distribution is what is usually meant, although several others are possible. In a bivariate normal distribution X is distributed normally, and so is Y. Also, and this is the nub of the matter, for any given value of X, the conditional values for Y are normally distributed. Likewise, for any given value of Y, the conditional values for X are normally distributed.

In most correlation research, a bivariate normal distribution is assumed. This assumption allows inferences about correlations to be extended to inferences about the *independence or the *dependence of the two variables. See *probability distribution, *normal distribution.

Bivariate Regression Coefficient   A regression coefficient showing the relation between two variables only. The coefficient indicates the degree of relationship between two variables by estimating the difference in the *dependent (*outcome) variable associated with a one-unit change in the *independent (*predictor) variable. The coefficient could estimate, for example, how much weight varies on average given a 1-inch or 1-centimeter increase in height. A *standardized bivariate regression coefficient is a *Pearson's r. See *simple regression, *regression equation. Compare *multiple regression analysis.

Biweight Mean   A measure of *central tendency designed to correct for extreme values by letting the researcher treat *outliers differently (that is, give them less weight) than other values. See *trimmed mean.

Black Box   Any mechanism whose internal workings are hidden. Said of input-output *research designs where what happens in between is impossible to study or is ignored.

  For example, studies of the effects of education on students often treat schooling as a black box. Students' characteristics upon entering and upon leaving are compared, but without directly considering which parts of the school experience might have produced any changes or how they could have done so.

Blank Experiment   An experimental *control produced by introducing an irrelevant treatment from time to time to keep subjects from becoming automatic in their responses.

Blind Analysis   Diagnosis or analysis made from test or experimental data without direct contact with or knowledge of the subjects, or when the person doing the analysis does not know which *treatment, if any, the various subjects received. The goal of such "blinding" is to eliminate bias. Ideally, neither the person administering the treatment, nor the person receiving it, nor the person analyzing the results should know who received the treatment, an alternative treatment, or a *placebo. Compare *double-blind procedure.

Block   (a) In *experimental design, a group of similar subjects receiving treatments; blocking in experiments is equivalent to *stratifying in surveys. See *randomized-block design. (b) In *path or regression analysis, a group of variables (e.g., *background variables) that can be put in a causal order even while the individual variables in the group or block cannot be so ordered. See *recursive model.

Block Design   An experimental design in which subjects are grouped into categories or "blocks." These blocks may then be treated as the experiment's *unit of analysis. The goal of categorizing subjects into blocks is to *control for a *covariate. See *matched pairs and *randomized block design for an example.

Block Sampling   (a) A sampling design in which respondents are grouped into representative categories or "blocks," which are then sampled.

(b) Another term for *area sampling. Compare *cluster sampling, *stratified sampling.

BLUE   *Best Linear Unbiased Estimator. A *regression line computed using the *least squares criterion when none of the *assumptions is violated. Estimates so made will be more accurate, on average, than estimates using other criteria.

BLS   Bureau of Labor Statistics.

BMDP   Biomedical Data Package. A statistical package containing several software programs. Compare *SAS, *SPSS.

BMI   *Body Mass Index.

Body Mass Index   An individual's weight (in kilograms) divided by the square of his or her height (in meters). Sometimes called the Quetelet index after its 19th-century inventor.

Bogardus Social Distance Scale   An attitude scale for measuring how closely people would be willing to associate with members of social and ethnic groups other than their own. Respondents are asked whether they would be willing, for example, to live in the same town as, in the same neighborhood as, invite home for dinner, or have a relative marry a member of the social group in question. Named after its creator, Emory S. Bogardus.

Boilerplate   Chunks of text (or other data) that are used repeatedly, word for word, in different documents. The term comes originally from old newspaper printing technology.

Bonferroni Technique (or Test or Inequality)   A method for testing the *statistical significance of *multiple comparisons (such as a series of *t tests of the means of three or more groups). It involves adjusting the significance level needed to reject the *null hypothesis by dividing the *alpha level you want to use by the number of comparisons you are making. Using the Bonferroni technique helps the researcher avoid the increased risk of *Type I error that comes with multiple comparisons. Also called "Dunn's Multiple Comparison Test." See *omnibus test.

For example, if a researcher wanted to use an alpha level of .05 and planned to make six comparisons, the new alpha level would be .05/6 = .008.

A researcher using the so-called "pseudo-Bonferroni technique" uses a more rigorous alpha level but does so without actually calculating the precise level. This more casual practice is quite common. Researchers will often say something like "because we made six comparisons, we used a more demanding significance level (.01 rather than .05)." Using the true Bonferroni technique, the researchers would have to conclude that their .01 level was not quite demanding enough.

Boolean Algebra   A form of algebra that deals with logical relations rather than numbers; or, a form of symbolic logic similar to algebra. Boolean algebra is important in computer design, set theory, and *probability theory. Named after the English mathematician, George Boole (1815–1864).

Boot (or Boot Up)   To start a computer; to get it ready to work by loading the *operating system into its memory. Derived from the term "bootstrap," as in "pull yourself up by your own bootstraps," meaning to get yourself going.

Bootstrap Methods   Procedures that provide alternative ways to estimate *standard errors by repeated *resampling from a sample. Bootstrapping is a *nonparametric approach to *statistical inference that can be applied to any data because it requires no assumptions about *underlying population distributions. See *jackknife method.

   The phrase "pull yourself up by your own bootstraps," which tells you to rely on your own resources, is apt for these methods. The researcher's own resources are the sample. Rather than make assumptions about *underlying population distributions to estimate the standard error, one estimates on the basis of repeated random samples (with replacement) from one's sample. Taking 100 such subsamples is common, and 1,000 is not unusual. This resampling provides an estimate of what we *would have* gotten had we sampled repeatedly from the *population*. Since resampling from the population is almost never done, bootstrapping is a more empirical technique.

Bounded Rationality   Limits (bounds) on the cognitive powers of decision makers, whether managers, customers, or researchers. These limits make optimal decisions rare in principle as well as in practice.

Bounds   Short for "confidence bounds," another term for *confidence limits.

Box-and-Whisker Diagram   A type of graph in which boxes and lines show a *distribution's shape, *central tendency, and *variability. The "boxplot," as it is often called, gives an informative picture of the values of a single variable and is helpful for indicating whether a distribution is *skewed and has *outliers. See *exploratory data analysis.

   In the following example, two box-and-whisker diagrams are used for comparing two distributions. The grades of two groups of students on the same 60-item test are diagrammed. Here is some of the information necessary to interpret the diagrams. (Terms and symbols vary, but the following conventions are fairly common and illustrate the main concepts.)

1. The upper and lower boundaries of each box (called *hinges) are drawn at the 75th and 25th *percentiles; this means that the box represents the *interquartile range (IQR), that is, the middle 50% of the values in the distribution.

B

2. The heavy line in the box (sometimes marked with an asterisk) shows the distribution's median.

3. The "whiskers" are the lines extending from the boxes. They reach to the largest and smallest scores that are less than 1 interquartile range (IQR) from the ends of the boxes.

4. Any points beyond the high and low points of the whiskers are *outliers (if they are less than 1.5 IQRs from the end of the box) and are marked with an "O." If they are more than 1.5 IQRs from the end, they are extreme outliers and are indicated by an "E."

5. Comparing the two boxplots, we can see that the variability in Group 2 is greater than it is in Group 1. Group 1's median score is lower than Group 2's. This is true despite the fact that the highest single score was earned by a student in Group 1 (the outlier, O1) and the lowest scores were earned by students in Group 2 (as is shown by its whisker).

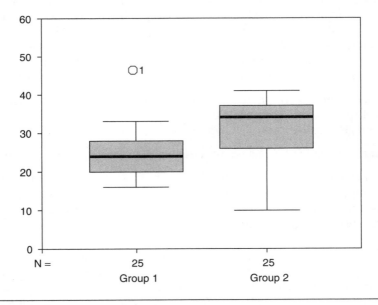

Figure B.3   Box-and-Whisker Diagram

Box-Cox Transformation   A method of transforming a continuous *dependent variable so that it does not violate the assumptions of statistical tests, especially *normality.

Box-Jenkins Methods   Techniques for *time-series problems, particularly *forecasting economic trends; they are one type of *ARIMA model.

Boxplot   Another term for *box-and-whisker diagram.

Box's Test   A test for *equality of variances in different populations.

Box-Score Method   An elementary first step in synthesizing research in *meta-analysis, named after its similarity to scorekeeping in baseball. Basically, one simply tallies the various research reports in terms of whether each supports or fails to support the hypothesis being studied.

Bracketing   (a) Another term for *collapsing data. (b) Providing upper and lower limits for a quantity. (c) In *qualitative research, considering something apart from its context.

Brainstorming   A method of problem solving that involves getting together a group of five to ten persons to consider a problem. The participants are to suggest whatever idea pops into their heads, no matter how weird, in the hope that a creative solution will be found. Compare *focus groups.

Buffer   (a) In brains and computers, a place for (or a process of) storing information briefly until one has time to deal with it. (b) Unscored test items included to reduce interaction between other items. Compare *blank experiment.

Bump Hunting   Looking at *spikes and *modes in frequency distributions that might indicate that some of the subjects are from different populations.

Byte   A unit of information used by digital computers, usually equal to 8 *bits. For example, in *ASCII the uppercase letter B is symbolized by the byte 01000010; C is 01000011.

C   (a) A *programming language developed at Bell Laboratories. Because of its efficiency and because Bell Labs was barred from copyrighting the program, it is very widely used, especially by professional programmers. C⁺⁺ is an advanced version. Compare *BASIC. (b) Symbol for *Pearson's contingency coefficient. (c) As a *superscript, symbol for *complement of; for example, $P(B^c)$ means the probability of the complement of B.

CAD   Computer-Assisted Design.

CAL   Computer-Assisted Learning.

Calculus   A branch of mathematics that analyzes rates of change, motion, and the slopes of curves, including density curves such as the *normal distribution. Its importance to statistics is that its formulas indicate how much the *dependent variable changes given a very small change in the *independent variable.

Call Back   Returning to *respondents in *survey research who were unavailable when the investigator first tried to contact them. Repeated callbacks can help reduce *nonresponse bias.

Campbell Collaboration   An association of researchers working to establish an *archive of *meta-analyses of research in the social and behavioral sciences. See *Cochrane Collaboration.

Canon   (a) Accepted rules or standards. Compare *algorithm. (b) An authoritative list of books. In research *design and statistics, most lists of canonical works would probably include volumes by Ronald Fisher and Donald Campbell.

C

Canonical Analyses   Any of several methods for studying relations among sets of related variables, including *multiple regression analysis, *discriminant analysis, *MANOVA, and *canonical correlation analysis.

Canonical Correlation Analysis   A form of *regression analysis for use with two or more independent variables and two or more dependent variables. The independent and dependent variables are each grouped into *linear composites or sets of variables; then correlations between those composites are calculated. Like other correlation coefficients, canonical coefficients range from −1.0 to +1.0. It is symbolized R_c. Today, canonical correlations are increasingly superseded by *structural equation models (SEM). Compare *factor analysis, *principal components analysis, and *MANOVA.

For example, suppose researchers were interested in the relation of students' health to their school achievement. They might want to use several measures of health (such as number of absences due to illness, nutritional information, body weight, dental records, school nurse's evaluation) and several measures of achievement (grades, scores on a reading test, scores on a mathematics test, teacher evaluation, and so on). The relation of the two clusters of measures could be studied with canonical correlations.

Canonical Variate   What sets of variables are called in *canonical correlation analysis. Compare *factor.

Cap   In *set theory, the symbol ∩, meaning "and." It is used to indicate the *intersection of two sets. Compare *cup.

Capital   Productive wealth; resources one can use to generate income or additional resources. Compare *cultural, *human, and *social capital.

Examples of economic capital include the balance in an interest-bearing savings account or tools one could use to make products to sell. By contrast, human capital refers to individuals' knowledge or skills; social capital means resources arising from social interaction; cultural capital refers to benefits arising from social standing.

CAQDAS   Computer-Assisted Qualitative Data Analysis Software. Software packages for assisting researchers in the *coding and sorting of qualitative data. Such software is used most often with text, such as interview transcripts. As with software for quantitative data analysis, the efficiency of qualitative software makes analyses that were once overwhelmingly time-consuming comparatively easy. It helps researchers analyzing large amounts of text to be more systematic and less impressionistic. Of course, no software, whether for qualitative or quantitative data, replaces intelligence and judgment in the interpretation of its output.

Carryover Effects   Lingering effects of an earlier experimental treatment that combine with the effects of a later treatment in a way that makes it difficult to assess the unique effects of the later treatment. Carryover effects are often a problem in *within-subject designs. See *practice effects, *A-B-A-B designs, *counterbalancing.

CARS   Computer-Assisted Reference Service. Located at major research libraries in the United States, CARS enables users to search for bibliographic citations by topic in a large *database.

CART   Classification and Regression Tree. See *classification tree.

Cartesian Coordinates   The numbers associated with points on a graph. The graphs one typically sees in the social and behavioral sciences display only the upper right-hand corner of the Cartesian chart (the positive numbers only). Named after the French mathematician René Descartes (1596–1650).

In the following figure, five coordinates are plotted. Reading clockwise from the upper right they are: (1) Y = 2, X = 3; (2) Y = −1, X = 4; (3) Y = −2, X = 4; (4) Y = −3, X = −3; (5) Y = 4, X = −3. For example, starting from the intersection at zero of the X and Y axes, for Y = 2, X = 3, you would move two spaces up and three to the right.

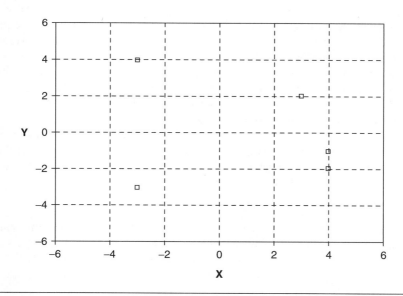

Figure C.1   Cartesian Coordinates

Cartesian Product   All the possible pairs of two *sets.

For example, if the two sets were two dice, the Cartesian product would be the 36 possible combinations of those two sets (dice)—1 + 1, 1 + 2, 1 + 3, and so on. See *underlying distribution.

Case   A *subject, whether an individual person or not, from which *data are gathered. A case is the smallest unit from which the researcher collects data. It is represented by a row (or *record) in a statistical program. The *sample size of a study equals its number of cases. Compare *unit of analysis.

Case-Cohort Design   A research design in which a random sample is drawn from a *cohort; the sample is used as a *comparison group for all the cases that occur in the cohort. Compare *resampling.

Case-Control Study   A method of studying an outcome by sampling cases with and without that outcome and studying their backgrounds. For example, in a study of lung cancer, the cases are individuals who have the disease. The controls are similar people who do not have it. The backgrounds of those with and without the disease are compared to understand the origins of the disease. Developed in *epidemiology, the method has many potential applications in the social sciences.

Case-Referent Study   Another term for *case-control study.

Case-Study Method   Gathering and analyzing data about one or a small number of examples as a way of studying a broader phenomenon. This is done on the assumption that the example (the "case") is in some way typical of the broader phenomenon. The case may be an individual, a city, an event, a society, or any other possible object of analysis. A wide variety of methods may be used to study the cases. An advantage of the case-study method is that it allows more intensive analyses of specific empirical details. A disadvantage is that it is hard to use the results to generalize to other cases. See *abstraction, *comparative method, *generalizability.

For example, a political scientist wishing to study why some candidates for public office are successful and others are not might study a particular election campaign in great depth in the hope of finding some general lessons about the electoral process.

Casewise Deletion   See *listwise deletion.

Catastrophe Theory   A theory of how living systems grow and differentiate. Long periods of slow change are punctuated by dramatic ("catastrophic") change. Originally developed in biology in the 1970s by René Thom, it has been applied to sociological, linguistic, and economic change as well.

Categorical Data Analysis   Any of several methods used when the variables, especially the *dependent variables, to be analyzed are categorical rather than continuous (measured on an *interval or *ratio scale). These include the *chi-squared test, *log-linear analyses, *logistic regression, and *probit regression.

Categorical Variable A variable that distinguishes among subjects by sorting them into a limited number of categories, indicating type or kind, as religion can be categorized: Buddhist, Christian, Jewish, Muslim, Other, None. Breaking a continuous variable, such as age, to make it categorical is a common practice, but since this involves discarding information it is usually not a good idea. Also called "discrete" or "nominal" variable. Compare *attribute, *continuous variable, *nominal scale.

Catell's Scree Test A graphic way to decide upon the number of important factors in a *factor analysis. See *scree plot for an illustration.

CATI Computer-Assisted Telephone Interviewing

Causal Conclusion A conclusion drawn from a study designed in such a way that it is legitimate to infer *cause. Most people who use the term "causal conclusion" believe that an experiment, in which subjects are *randomly assigned to *control and *experimental groups, is the only *design from which researchers can properly infer cause. Other researchers, such as those who do *path analysis on nonexperimental data, do not agree with that restriction. Compare *correlational research design, *natural experiment, *ecological fallacy.

Causal Diagram A graphic representation of *cause and *effect relations among *variables. Arrows indicate the direction of causal influence. See *model and, for an example, *path diagram.

Cause An event, such as a change in one variable, that produces another event, such as a change in a second variable. See *necessary condition, *sufficient condition.

  Before reading further, be forewarned. There is no concept in this dictionary more troublesome than "cause." Highly respected researchers disagree about what constitutes a cause, and especially about how restrictive a set of conditions must be met before it is legitimate to talk of cause. Many social scientists—and even some philosophers—would agree with the following. Others, naturally, would not.

  To attribute cause, for X to cause Y, three conditions are necessary (but not sufficient): (1) X must precede Y; (2) X and Y must covary; (3) no rival explanations account as well for the covariance of X and Y.

  Causal relations may be simple or multiple. In simple causation, whenever the first event (the cause) happens, the second (the effect) always does too. Multiple causation is much more common in the social and behavioral sciences. Multiple causes may be such that any one of several causes can produce the same effect. For example, monetary inflation may be caused by rising wages, rising prices, declining productivity, or some combination of the three. Multiple causes may also be such that no one of them will

C

necessarily produce the effect, but several of them in combination make it more likely. For example, prejudiced attitudes may be produced by repressive child rearing, general ignorance, low self-esteem, and/or lack of contact with people different from oneself.

Ceiling Effect   A term used to describe what happens when many *subjects in a study have scores on a *variable that are at or near the possible upper limit ("ceiling"). The ceiling effect makes analysis difficult, since it reduces the amount of variation in a variable. Compare *floor effect.

For example, suppose a group of statistics professors wants to see whether a new method of teaching increases knowledge of elementary statistics. They give students in their classes a test, try the new method, and then give the students another version of the same test to see whether their scores went up. If one of the professors had students who knew a lot of statistics already, and scored at or near 100% on the first test, she could not tell whether the new method was effective in her class. The scores of her students were so high (at the ceiling) that they could hardly go up, even if the students learned a great deal using the new method.

Cell   (a) The space formed by the intersection of a row and a column in a statistical table. (b) Any single group in an *analysis of variance design.

For example, in the table below, each of the numbers (except for the 100s) is in one of the table's cells. The percentage of Seniors who plan to Seek Work is in one cell, as is the percentage of Juniors who Don't Know, of Sophomores who plan to go on to Grad. School, and so on.

Table C.1   Cell: Students' Plans Upon Graduation, by Class

	Freshmen	Sophomores	Juniors	Seniors
Seek Work	73	58	57	58
Grad. School	5	11	18	30
Don't Know	22	31	25	12
Total	100	100	100	100

NOTE: Numbers listed are percentages.

Censored Data   Data that are incomplete in some way, as when certain values, usually extreme values, are unknown or ignored. Sometimes used as a synonym for truncation; when a distinction is made, censored usually refers to *measurements* that are incomplete, while truncation refers to incomplete *samples*. See *trimmed mean, *truncated distribution.

Imagine, for example, a study conducted in 2005 of the college graduation rate of individuals born in 1980. Some people born in 1980 may not have completed college by 2005, but they could easily do so later on, after the end of the study. The data in this study would be censored, that is, the number of people from the 1980 *cohort who got their degrees after 2005 would be unknown.

Censored Regression Model   See *tobit analysis.

Census   A complete count of an entire *population by observing each of its members—in contrast to a survey of a *sample of the population.

Census Tract   A small area of a city or other densely populated region in the United States defined to make gathering census data easier. Census tracts usually contain between 3,000 and 8,000 people (mean, 4,000). They are defined with the advice of local committees so as to approximate neighborhoods.

Centile   Abbreviation of *percentile.

Central Limit Theorem   A statistical proposition to the effect that the larger a sample size, the more closely a *sampling distribution of a statistic will approach a *normal distribution. This is true even if the population from which the sample is drawn is not normally distributed. A sample size of 30 or more will usually result in a sampling distribution for the mean that is very close to a normal distribution.

The central limit theorem explains why *sampling error is smaller with a large sample than with a small sample and why we can use the *normal distribution to study a wide variety of statistical problems.

Central Tendency, Measure of   Any of several statistical summaries that, in a single number, represent the typical number in a group of several numbers. Examples include the *mean, *mode, and *median. Compare *dispersion, *variability.

A batting average is a well-known measure of central tendency in the United States. A grade point average might be a more important example for many college students.

Central Tendency (of a Distribution)   A point in a distribution of scores that corresponds to a typical, representative, or middle score in that distribution—such as the *mode, *mean, and *median.

Centroid   (a) A *weighted combination of the observed *dependent variables in a *MANOVA; it is a mean of the *vector scores for all subjects. (b) In *factor analysis, the "centroid method" is a way to extract factors. (c) The mean *discriminant function score for each group in a *discriminant analysis.

CERES Plot   A graphic means of judging whether the *independent (*explanatory, *predictor) variables in a *regression analysis have a linear relation to the *dependent (*outcome, *response) variable. CERES stands for "combining conditional expectations and residuals."

Ceteris Paribus   Latin phrase meaning "other things being equal"; also often used to mean "if other things remain unchanged." The phrase is generally used to qualify a conclusion, as "this is true *ceteris paribus,*" meaning this is true if other things are/remain equal/unchanged.

Many aspects of *research design (such as *random assignment to *experimental groups or statistically *controlling for a variable) can be seen as attempts to approach the goal of *ceteris paribus.*

CF   *Cumulative Frequency.

CFA   *Confirmatory Factor Analysis.

Chain   A series of values in which a value at one point depends in some way on the previous values in the series.

Chain Path Model   Said of measurements of a variable taken from the same sample at three or more different times when the cause of the value of the measurement for each time is the immediately previous measurement. Also called *Markov chain.

For example, say we took measurements of some variable at four times, T1 → T2 → T3 → T4. Using a chain path model, we would assume the cause of the value at time 4 (T4) is T3, but T2 and T1 have no effect. Similarly, the cause of T3 is T2, but T1, while causing T2, has no direct effect on T3.

Chance Error   Another term for *random error. Compare *noise.

Chance Variable   Another term for *random variable and *stochastic variable; also called *variate.

Chance Variation   Another term for *random variation.

Change Score   A score obtained by subtracting a *pretest score from a *posttest score. A more common approach, and one that can reduce measurement error, is to treat the posttest score as a *dependent variable in a *regression analysis and the pretest score as *one of* the *predictor variables. Also called *difference score and "gain score."

Chaos Theory   An umbrella term for several methods of analyzing extremely complex systems, especially systems with many dynamic, *non-linear, and *random elements, such as the weather. Small errors in assumptions or small differences in initial conditions can throw predictions off by a wide margin. The "chaos" is never complete in chaos theory; behavior of the system is random, but recurring; causes exist, but they are difficult to discover and model with conventional methods and *linear equations. See *fractal, *stochastic, and *complexity theory.

Characteristic Root (or Value or Number)   Other terms for *eigenvalue.

Chebyshev's Theorem   Alternate spelling of *Tchebechev's theorem. (There are others, e.g., Chebycheff and Tchebycheff.)

Chi-Squared Distribution   A family of theoretical *probability distributions, defined in part by their different *degrees of freedom *(df)*. The *mean of the curve is its *df* and the *standard deviation is the *square root of $2 \times df$. The *chi-squared test is based on it.

Chi-Squared Test   A *test statistic for categorical data. As a test statistic it is used as a test of *independence, but it is also used as a *goodness-of-fit test. The chi-squared test statistic can be converted into one of several measures of association, including the *phi coefficient, the *contingency coefficient, and *Cramer's *V*.

   The chi-squared test is known by many names: Pearson chi-square, X^2, chi^2, and c^2.

   The simplest use of the chi-squared test, illustrated in the following example, occurs when a researcher wants to see if there are statistically significant differences between the observed (or actual) frequencies and the expected (or hypothesized, given the *null hypothesis) frequencies of variables presented in a *cross tabulation or *contingency table. The larger the observed frequency is in comparison with the expected frequency, the larger the chi-squared statistic. The larger the chi-squared statistic, the less likely the difference is due to chance, that is, the more statistically significant it is.

   For example, say that a researcher gives a pass/fail test to a sample of 100 subjects, 42 men and 58 women; 61 subjects pass, and 39 fail. If the researcher were interested in whether there are differences in test performance by gender, she could use the chi-squared test to test the *null hypothesis of no statistically significant differences between the sexes. To do so, she might arrange the information about her subjects in Tables C.2, C.3, and C.4. Table C.2 gives the total (or *marginal) frequencies for the two variables. Table C.3 shows what the (approximate) frequencies of passes and fails on the two tests *would have* been if the null hypothesis were true, that is, what you would expect if there were no differences on the exam between men and women. Table C.4 shows the actual or observed number of men and women who passed or failed the exam.

Table C.2   Chi-Squared Test (Marginal Frequencies)

	Pass	*Fail*	*Total*
Men			42
Women			58
Total	61	39	100

Table C.3   Chi-Squared Test (Expected Frequencies)

	Pass	*Fail*	*Total*
Men	26	16	42
Women	35	23	58
Total	61	39	100

Table C.4   Chi-Squared Test (Observed Frequencies)

	Pass	*Fail*	*Total*
Men	19	23	42
Women	42	16	58
Total	61	39	100

Comparing Table C.3 and Table C.4, it is clear that the actual and expected frequencies are not identical. For example, 26 men were expected to pass, but only 19 did; 23 women were expected to fail, but only 16 did, and so on. But are these differences statistically significant, that is, are they unlikely to have occurred by chance? Conducting the chi-squared test can tell you. The answer (calculations not shown) is that the null hypothesis of no difference between men and women should be rejected. The differences are greater than what would be expected by chance alone; they are significant at the .01 level.

Chow Test   A *test statistic to determine whether *regression equations differ significantly among themselves or to ascertain whether regression coefficients change over time. It is based on the *F test.

CI   *Confidence interval.

Circular Reasoning   A kind of fallacy that occurs when one conclusion depends on a second, which in turn depends on the first. Compare *tautology.

    For example, "Unemployed people are lazy. How do we know this? Because they are unemployed, which proves they're lazy; if they weren't lazy they'd have jobs."

Class Boundary   See *class limits.

Class Frequency   The number of observations of a particular *variable that fall in a given *class interval.

    For example, if researchers were studying income distribution in a particular city and 2,149 individuals earned between $30,000 and $39,999, the class frequency for the class interval $30,000–$39,999 would be 2,149.

Classical Statistical Inference   What most people mean by "statistical inference." The word "classical" is often added to make a contrast with *Bayesian inference.

Classification Tree   Methods of predicting the category of an object from the values of its predictor variables. Classification trees are often used in *data mining. Also called "classification and regression trees" or CART. See *decision tree and *tree diagram for examples.

Classification Variables   Another term for *background variables.

Classificatory Variable   A *categorical variable, that is, one that values a variable by classifying or categorizing—such as upper/middle/lower class or jumbo/large/medium/tiny shrimp. Compare *nominal and *discrete variables.

Class Interval   A convenient grouping of the data on a *continuous variable that makes it easier to analyze; the interval between the boundaries (or limits) of a class, such as between 21 and 40 million in the following example. By turning continuous variables into *categorical variables, class intervals make it possible to do *frequency distributions and *cross tabulations—at the cost, however, of discarding detailed information.

For example, the following table classifies 236 countries with indigenous inhabitants by population size. This makes it easier to see the big picture, easier than it would be if we used a list of all the nations and their exact populations. On the other hand, a major disadvantage of using classes is that it obscures large differences, such as the one between Mauritius, with a population of about 1 million, and Australia, with more than 19 million, both of which are grouped together in the same category.

Table C.5   Class Interval: Distribution of Nations by Population Size (2003)

Population (in Millions)	Number of Countries
Less than 1	81
1–20	106
21–40	20
41–60	10
61–80	4
81–100	4
100+	11

Class Limits   The upper and lower values of a *class interval. Also called "class boundaries."

Clinical Significance   *Practical significance in a clinical setting. Usually contrasted with *statistical significance.

C

Clinical Trial   An *experiment comparing the effect of treatments. Originally designed for medical research, especially drug testing, clinical trials are sometimes held to be the "*gold standard" in other fields, such as education. See *double-blind procedure, *randomized control trial.

Cliometrics   The application of statistical methods to the study of history. Named after Clio, the muse of history. Compare *econometrics.

Closed Question Format   Also known as closed-ended questions. In surveys and interviews, researchers most often offer subjects a limited number of predetermined responses to questions (closed format) rather than allow them to choose their own words for answering questions (*open question format).

For example, "What sort of job is the president doing overall? (a) excellent (b) good (c) fair (d) poor (e) don't know." Using the closed-question format means that a respondent who wants to say "very good for foreign policy, but not so hot on domestic issues" is forced to select among options (a) to (e).

Closed System   A theoretical system that does not admit evidence or arguments from different perspectives. In other terms, a causal system that allows no *exogenous causal variables.

Freudianism, Marxism, and *behaviorism have been accused of being closed systems. Indeed, most theoretical systems have been so accused— by opponents.

Cluster Analysis   Any of several procedures in *multivariate analysis designed to determine whether individuals, cases, or other units of analysis are similar enough to be grouped into clusters. The individuals within a cluster are similar on some variable(s), while the clusters are dissimilar from one another. See *MANOVA, *canonical correlation, *discriminant analysis.

Cluster Randomizing   Assigning groups rather than individuals to a study's *control and *experimental conditions. For example, one might assign classrooms rather than individual students to receive or not receive a treatment. Compare *cluster sampling, *group randomized trial.

Cluster Sampling   A method for drawing a *sample from a *population in two or more stages. It is typically used when researchers cannot get a complete list of the members of a population they wish to study, but can get a complete list of groups or clusters in the population. It is also used when a *random sample would produce a list of subjects so widely scattered that surveying them would be prohibitively expensive. Generally, the researcher wishes to use clusters containing subjects as diverse as possible. By contrast, in *stratified sampling the goal is often to find strata containing subjects as similar to one another as possible.

C

The disadvantage of cluster sampling is that each stage of the process increases *sampling error. The margin of error is therefore larger in cluster sampling than in simple or stratified random sampling; but, since cluster sampling is usually much easier (cheaper), this error can be compensated for by increasing the sample size. See *central limit theorem.

For example, suppose you wanted to survey undergraduates on social and political issues. There is no complete list of all college students. But there are complete lists of all 3,000+ colleges in the country. You could begin by getting such a list of colleges (which are "clusters" of students). You could then select a *probability sample of, say, 100 colleges. Once the clusters (colleges) were identified, the researchers could go to each school and get a list of its students; students to be surveyed would be selected (perhaps by simple *random sampling) from each of these lists.

COBOL   Common Business Oriented Language. A *programming language. Compare *BASIC, *C, *FORTRAN.

Cochran's C Test   One of several tests for *equality of variances. Another is *Levene's test.

Cochran's Q Test   A variant of the *chi-squared test used for samples that are not *independent, that is, for *within-subjects designs. It is an extension of *McNemar's test to research problems with three or more *correlated samples.

Cochrane Collaboration   A group of medical researchers who conducts systematic reviews (*meta-analyses) of the research literature in medicine and make them available both to the public and to other researchers and medical practitioners. It is the model for a similar group in the social sciences, the *Campbell Collaboration.

Code   (a) Rules specifying how data are to be represented. See *code book. (b) Rules for converting data from one form to another. See *coding. (c) A *computer program, as in, "She wrote the code for that operating system."

Code Book   A list of the *variables and how they have been coded so that they can be read and manipulated by a computer. Also called a "coding frame." See *coding.

For example, a typical entry in a code book would be: Variable 1, Sex: 1 = female; 0 = male.

Coding   (a) "Translating" data from one language or format into another—often to make it possible for a computer to operate on the data thus coded. Some coding schemes are decided before data are collected, especially when the data are easily defined and measured. By contrast, in *qualitative research coding is often done after the data are collected. A familiar example would be examining transcripts of interviews in order to devise a way of coding them. Computer software packages are often used for this kind of coding. See *effects, *dummy, and *contrast coding. (b) Writing a set of instructions telling a computer how to handle data. See *programming.

For example, (a) if one of your variables were "race" you might code these as 1 for "black" and 2 for "white," and 3 for "other." (b) "Arrange the data in ascending numerical order" could be coding that instructed a computer about how to handle a data file.

Coefficient   (a) A number used as a measure of a property or characteristic. (b) In an equation, a number by which a variable is multiplied.

To find a specific coefficient in this dictionary, see under the type or kind; for instance, to find coefficient of determination, look under "determination, coefficient of."

For example, (definition a) a coefficient inequality between incomes could be calculated by dividing the larger income into the smaller. Thus, if the average (mean) income of U.S. men working full time were $25,000 per year, and that of women working full time were $15,000, the coefficient of inequality between men and women would be $15,000/25,000 = .6$. In the equation $Y' = a + 3.2X + e$, 3.2 is a coefficient (definition b). See *regression coefficient.

Cofactor   A variable that has a particular effect in combination with another variable.

For example, some infectious diseases, such as hepatitis B, increase susceptibility to noninfectious diseases, such as liver cancer. Compare *interaction effect.

Cognitive Science   The interdisciplinary study of cognition, that is, the processes of acquiring, creating, and disseminating knowledge. It is composed in varying proportions of cognitive psychology, computer science (*information theory), philosophy (*epistemology), and linguistics.

Cohen's *d*   Standardized mean difference *effect size measure. See effect size, definition (b).

Cohen's Kappa   A measure of *interrater reliability for *categorical data. This percentage-of-agreement measure corrects for chance or random agreement. Kappa is 1.0 when agreement is perfect; it is 0.0 when agreement is no better than would be expected by chance. See *reliability coefficient.

Coherence   A measure of the strength of association of two *time series.

Coherency Principle   In *Bayesian analysis, the belief that *subjective probabilities, often described as betting odds, follow the usual laws of probability.

Cohort   A group of individuals having a statistical factor (usually age) in common. Compare *social category.

For example, all persons born in 1981 form a cohort.

Cohort Analysis   Studying the same *cohort over time. See *panel study, *time-series analysis.

For example, individuals who graduated from high school in 1985 form a cohort whose subsequent educational experiences could be followed in a cohort analysis—for example, how many went on to college immediately? how many went on eventually? of those who attended college, how many went to a two-year college? how many to a four-year college? how many graduated?—and so on.

Cohort Effects   The effects of membership in a cohort, usually an age group. Also called "generation effects." Often contrasted with *period effects, which are the effects of living during an era regardless of individuals' ages.

For example, it is often claimed that being an adolescent in the 1950s affected people's attitudes quite differently than being an adolescent in the 1960s.

Cohort-Sequential Design   A combination of *cross-sectional and *longitudinal designs, usually to test for generation or *cohort effects. Compare *accelerated longitudinal design.

For example, in the following table the rows allow longitudinal comparisons, while the columns provide for cross-sectional analysis. The *cells contain the ages of the subjects.

Table C.6   Cohort-Sequential Design: Ages and Dates of Measurement of Subjects

	Year of Measurement				
Birth Year	1985	1990	1995	2000	2005
1980	5	10	15	20	25
1985		5	10	15	20
1990			5	10	15
1995				5	10
2000					5

Cohort Study   A study of the same group (cohort) over time, but not necessarily of the same individual members of that group. Contrast *panel study, in which the same individuals are studied at each stage.

For example, you might draw *probability samples from the cohort of 1975 (those born in that year) in the first three presidential election years in which they were eligible to vote: 1996, 2000, and 2004. You would be unlikely to study the same individuals more than once.

COLA   Cost of living adjustment; an increase in wages or benefits, usually based on an index of consumer prices. See *consumer price index.

Collapsing   Combining groups or categories of a variable in order to reduce their number. Also called "bracketing." See *class interval.

For example, suppose we surveyed 100 people about the number of movies they saw last year and got the following results. Table C.7 uses five categories. Table C.8 collapses the five categories into two.

Table C.7   Collapsing (Five Categories): Number of Movies Seen Last Year

0–5	22
6–10	18
11–15	12
16–20	28
20+	20

Table C.8   Collapsing (Two Categories): Number of Movies Seen Last Year

0–15	52
16+	48

Collinear   Having a common line. See *multicollinearity.

Collinearity   The extent to which *independent (or *predictor) variables in a *regression analysis are correlated with one another. Collinearity causes problems in analysis because it makes it difficult to study the separate effects of independent variables. The main difference between collinearity and *multicollinearity is two syllables. See *intercorrelation, *tolerance.

Column Marginals   See *marginal frequency distribution and *row marginals.

Combination   See *permutation.

Combinatorics   The study of the ways items in a *set can be arranged, selected, or combined. For example, if there were five candidates for two awards, such as first and second place, a total of 10 distinct combinations of winners are possible.

Commonality Analysis   Methods of separating the effects of correlated *predictor variables in *multiple regression analysis and other multivariate techniques. The problem addressed by commonality analysis is estimating how much of the variance in the *outcome variable can be explained by each of the predictors, while controlling for the other predictors, and how much can be explained by all the predictors taken together. See *R^2.

Common Factor A *factor that appears in two or more variables. A factor appearing in only one variable is called a specific factor. See *factor analysis.

Common Factor Variance The variance that two or more *factors share. See *communality, *factor analysis.

Common Logarithm A *logarithm using base 10.

Common Metric A scale of measurement shared by more than one study or into which the results of several studies have been transformed. *Transformation to a common metric is usually accomplished using *standard scores; this is frequently done in a *meta-analysis because it allows the results of different studies to be compared.

Common Variance Variance shared by two or more variables. Two measures of the same variable, such as mental ability, would be expected to have a good deal of common variance and thus to correlate highly. The variance not shared by the two measures would be the *error variance. See *reliability.

Communality The proportion of the total variance that is *common factor variance (that is, shared by two or more factors). It is calculated by summing the squared *factor loadings (see below) of a variable. Symbolized h^2.

Table C.9   Communality

	Factor 1	Factor 2
Loadings	.70	.10
Loadings Squared	.49	$.01 = .50 = h^2$

Communications Theory The study of the transfer of information. It tends to emphasize parallels between the ways humans and computers do this. Compare *information theory, *artificial intelligence, *cognitive science.

Comparative Method The study of more than one event, group, or society to isolate factors or *variables that explain patterns. The term is perhaps most often used to describe cross-national research. However, almost all systematic research is comparative in the broad sense of the term. Experimental research, for example, involves comparing *control and *experimental groups. The term "comparative method" is often used when the research involves secondary analysis of historical data. (When data are contemporary, the term "cross-cultural" is sometimes used instead.) Compare *natural experiment. For an example, see *abstraction.

There are two basic strategies in comparative research: (1) study events or groups that differ in many ways but that have one thing in common—for instance, different societies that have experienced revolutions; (2) study societies or groups that are highly similar, but differ in one important respect—such as modern, industrial nations that have different economic policies.

Comparison Group Another term for *control group; it is most often used when the control group receives an alternative *treatment rather than no treatment. The term is more common in *nonexperimental than *experimental research, but is used in both.

Competing Risks In *survival analysis, the subcategories of ways that a more general event (failure) can occur. For example, teachers may leave the school in which they teach for several reasons—to take another teaching job, to take a job in a different line of work, to become a stay-at-home parent, by retiring, by dying, and so on. To a district wishing to retain its teachers, each departure is a "failure" in survival analysis terms. The different reasons for or ways of departing are the competing risks (of losing a teacher). If the importance of one of the competing risks can be reduced, this will increase the average time to failure for all reasons combined.

Complement (of A) In *set theory, "not A." Also called "negation of A" and "complementary event."

For example, if a set is made up of the numbers from 1 to 10, and a subset "A" is 2, 5, 8, 9, and 10, the complement of A is the rest of the numbers in the set, that is, 1, 3, 4, 6, and 7.

Complete Case Analysis Analyzing only those cases with complete information on all variables. Compare *listwise deletion.

Completely Randomized Design An *experimental design in which *treatments are assigned to subjects at random.

Complex Comparison A comparison of two or more groups with another group. Comparing Group A to Group B would be a simple comparison. Comparing Group A plus Group B to Group C would be a complex comparison. Compare *pairwise comparison.

Complexity Theory The study of computer and mathematical models for analyzing topics with very large numbers of interacting variables, such as large corporations or government agencies. *Chaos theory is a variant. See *fuzzy set.

Component Bar Chart Another term for *segmented bar chart.

C

Components (of a Time Series)   The basic types of movement of *time-series data. There are four: *secular trend, short-term trend, cyclical variation, and *random variation.

Composite Score   A score made up of two or more scores, usually by adding the scores together or by computing their *mean.

Compound Bar Chart   See *segmented bar chart.

Computer Intensive Methods   Statistical methods that rely heavily on computers. While they might be theoretically possible without using computers, these methods are practically impossible without assistance from machines. As computers increase in power, "intensive" methods take less time. See *computer simulation and *resampling methods.

Computer Program   A set of instructions written in a form a computer can read ("machine readable") that tells it how to perform specific tasks.

Computer Simulation   Using a computer to build a *model of what would happen in a real-world situation under certain conditions. Computer simulations are used in a wide variety of fields, from economic forecasting to statistics. Compare *Monte Carlo methods.

Concentration, Coefficient of   Another term for *Gini coefficient.

Concentration Ratio   Any of several measures of the extent to which economic activity in an industry is concentrated in a small number of firms. If the activity were monopolized by one firm, the ratio would be 1; if it were controlled equally by 100 firms, it would be .01. Compare *Gini coefficient.

Concept   An abstract idea that enables one to categorize data. It often implies generalization from particulars—although Plato would not agree. Compare *construct, a term used more often in quantitative research to express the same idea (concept) as "concept."

For example, if you saw an unusual breed of dog for the first time, you would probably still recognize it as a dog—even though you had never seen it before—because it would fit into your general concept or idea of what a dog is. Compare *schema.

Conceptualization   Specifying what we mean by the *concepts we will use in a research project. This is often a step on the way to *operational definitions.

Concomitant Variable   A variable a researcher wishes to *control for. Often an *attribute or a *trait of subjects. Also called *covariate.

Concomitant Variation   Said of two or more phenomena that vary together, or covary. See *correlation, *covariance, *cause.

The term was introduced by J. S. Mill (1806–1873) to describe a method for determining a causal link between two phenomena. The procedure is to investigate whether, when a supposed cause is present, the effect is present and when the supposed cause is absent, the effect is absent.

Concordance, Coefficient of   See *Kendall's coefficient of concordance.

Concurrent Validity   A way of determining the *validity of a measure by seeing how well it correlates with (agrees or "concurs" with) some other measure the researcher believes is valid.

For example, if psychologists wanted to see whether a new IQ test were a valid measure of intelligence, they could correlate subjects' scores on the new test with their scores on an old IQ test that they thought was a good measure of intelligence. If the scores were highly correlated, this would be evidence of the validity of the new test—or, at least, that the two tests were measuring the same thing.

Condition   A *treatment or a *level of an *independent variable in an *experiment.

For example, a study comparing the effects of drugs A, B, and C has three conditions (Drug A, Drug B, and Drug C). The independent variable (drug treatment) has three levels (A, B, and C).

Conditional Distribution   The distribution of the values of one variable as it is influenced by the values of another variable or variables. See *bivariate distribution.

For example, the distribution of weight in a population is conditional upon the distribution of height. (Tall people tend to weigh more than short people.) If we compared the distribution of weights of people 5 feet tall with that of people 6 feet tall, we would be comparing conditional distributions. A *regression equation predicts the mean of the conditional distribution of the dependent variable for every value of the independent variable(s).

Conditional Effect   Another term for *interaction effect. It is used especially in *regression analysis to distinguish it from general effects; "interaction effect" is the more common term in the context of *ANOVA.

For example, going to college improves earnings for all groups (general effect), but the improvement is greater for some groups (conditional effect).

Conditional Event   In *probability theory, an event that can occur only in conjunction with another event. See *conditional probability. Contrast *independent event.

For example, say you rolled a pair of dice one at a time. Getting a total of nine (the conditional event) after the second roll is conditional upon the first die having come up three or higher.

Conditional Odds   *Odds that take into account other variables. Compare *conditional probability.

C

For example, the odds (unconditional) of graduating from high school in a particular state might be 80% to 20%, or 4 to 1. Taking into account the variable of sex, the conditional odds for females might be 85% to 15%, or 5.67 to 1.

Conditional Probability   (a) The chance that an event will occur, given that some other event has already occurred. Symbolized p(B|A), which is read, "the probability of event B, given event A." (b) The chance that one condition exists given that another does. See *Bayesian inference.

For example, (definition a) the probability of drawing an ace at random from a deck of 52 playing cards is 4 out of 52, or 1 out of 13. Say you drew a card, got an ace, and did not put it back in the deck. Call that draw event A. What is the conditional probability (given event A) of drawing another ace (event B)? Since you did not replace the first ace, the (conditional) probability of drawing a second ace is 3 out of 51, or 1 out of 17. Compare *gambler's fallacy.

An example of (b) might be the likelihood that a patient has the HIV virus, given that his blood test is positive. See *sensitivity.

Conditioning Effect   Another term for *interaction effect.

Confederate   Someone who pretends to be a subject in an *experiment, but who is actually helping the experimenter in some way.

Confidence Band   The region between the lower and upper *confidence limits.

Confidence Bounds   Another term for *confidence limits.

Confidence Coefficient   Another term for *confidence level. It is 1.0 minus the *alpha level. Thus an alpha level of .05 results in a confidence coefficient of .95.

Confidence Interval   A range of values of a *sample statistic that is likely (at a given level of probability, called a *confidence level) to contain a *population parameter. The interval that will include the population parameter a certain percentage (*confidence level) of the time. In other words, a range of values with a known probability of including the true population value. The wider the confidence interval, the higher the confidence level. See *confidence level for an example.

It is common to say, for example, that one can be 95% confident that the confidence interval contains the true value. Although this is the usual way to report confidence intervals and limits, it is not technically correct. Rather, it is correct to say, were one to take an infinite numbers of samples of the same size, that on average 95% of them would produce confidence intervals containing the true population value.

Confidence Level   A desired percentage of the scores (often 95% or 99%) that would fall within a certain range of *confidence limits. It is calculated

C

by subtracting the alpha level from 1 and multiplying the result times 100; e.g., $100 \times (1 - .05) = 95\%$.

For example, say a poll predicted that, if the election were held today, a candidate would win 60% of the vote. This prediction could be qualified by saying that the pollster was 95% certain (confidence *level*) that the prediction was accurate plus or minus 3% (confidence *interval*). The larger the sample, the narrower the confidence interval or margin of error.

Confidence Limits   The upper and lower values of a *confidence interval, that is, the values defining the range of a confidence interval. See *confidence level for an example.

Confidentiality   See *anonymity.

Confirmatory Factor Analysis   *Factor analysis conducted to test hypotheses (or confirm theories) about the factors one expects to find. It is a type of or element of *structural equation modeling. See *exploratory factor analysis.

Conflict Theory   A perspective on society and social relations contending that the main determinant of social phenomena is the tendency of individuals and groups to have opposing interests over which they come into conflict. Among the many classical authors who could be called conflict theorists, Karl Marx (1818–1883) and Max Weber (1864–1920) are probably the best known. Compare *functionalism.

Confound   (a) *verb:* To study combined treatments in such a way that their separate effects cannot be determined. (b) *noun:* A variable that obscures, or makes it impossible to interpret, the relations among other variables. See *confounded.

For example, to study the effects of fertilizer on your lawn when the fertilizer must be applied with water is to confound the effects of watering and of fertilizing. Water is the confound.

Confounded   Said of two or more *variables whose separate effects cannot be isolated.

For example, if Professor X used Textbook A in her class and Professor Y used Textbook B, and students in the two classes were given achievement tests to see how much they had learned, the *independent variables (the textbooks, and the professors' teaching effectiveness) would be confounded. There would be no way to tell whether any differences in achievement (the *dependent variable) between the two classes were caused by either or both of the independent variables.

Confounding Variable   A variable that obscures the effects of another. See *confound and *confounded for examples. Compare *suppressor variable.

C

Consensual Validation   The use of agreement (consensus) of two or more experts to determine whether a statement is true or valid.

Consequent   The second term in a ratio. In the ratio 3:2, 2 is the consequent. The first term (3) is the *antecedent.

Consequential Validity   Unintended negative consequences that arise from using a test or measurement, such as an adverse effect on a particular social group. The concept is discussed only to refer to *in*validity, not validity.

Conservative (Measure or Estimate)   Said of a statistic that tends to underestimate; that is, if it errs, it is more likely to do so by being overly cautious. See *post hoc tests.

For example, *omega squared is a conservative measure of *strength of association, because it is more likely to underestimate than to overestimate that strength—especially in comparison with *eta squared, which can sometimes overestimate the strength of an association.

Consistent Estimator   A sample *statistic (estimator) that tends to get closer to the true population *parameter as the sample size increases. See *central limit theorem.

Constant   (a) A measure or value that is the same for all units of analysis. (b) A quantity that does not change value in a particular context. (c) In a *regression equation, the *intercept (also called *regression constant and Y intercept) is often referred to as "the constant"; the *beta coefficients are also constants, but are less often so called. Compare *variable, *universal constant.

For example, (definition a) in research that studied variables explaining unemployment among women only, sex would be a constant; all subjects (units of analysis) are female. An example of (b) would be a price that does not change regardless of fluctuations in supply or demand. For definition (c), the value of a would be the constant in the regression equation $Y' = a + bX + e$.

Constant Comparison   A method of *qualitative data analysis particularly associated with *grounded theory. As data are being coded and analyzed, the researcher continually compares conclusions drawn from the earlier stages to data from the later stages. For example, initial review of some interview transcripts will suggest coding schemes and interpretations. With each additional transcript reviewed, these schemes and interpretations are compared and revised if needed.

Construct   (a) Something that exists theoretically, but is not directly observable. (b) A *concept developed (constructed) for describing relations among phenomena or for other research purposes. (c) A theoretical (not

C

*operational) definition in which concepts are defined in terms of other concepts.

For example, intelligence cannot be directly observed or measured; it is a construct. Researchers infer the existence of intelligence from behavior and use *indexes (such as size of vocabulary or the ability to remember strings of numbers) to "construct" a measure of the construct, intelligence.

Construct Validity   The extent to which *variables accurately measure the *constructs of interest. In other words: How well are the variables *operationalized? Do the *operations really get at the things you are trying to measure? How well can you generalize from your operations to your construct? In practice, construct validity is used to describe a *scale, *index, or other measure of a variable that *correlates with measures of other variables in ways that are predicted by, or make sense according to, a theory of how the variables are related. See *concurrent, *content, *convergent, and *criterion-related validity. Absolute distinctions among these kinds of validity are difficult to make, in large part because procedures for assessing them tend to be similar if not identical. *Convergent and *discriminant validity, for instance, are used as tests of construct validity.

For example, if you were studying racist attitudes, and you believed that racism (the construct) was more common among people with low self-esteem, you could put together some questions that you thought were a good *index of racism. If subjects' scores on that index were strongly (*negatively) correlated with their scores on a measure of self-esteem, this would be evidence that your index had construct validity. The index is more likely to be a good measure of racism if it correlates with something your theory says it should correlate with than if it does not. All this assumes, of course, that your theory is right in the first place about the relation between self-esteem and racism *and* that you have a valid measure of self-esteem.

Constructionism   A research perspective that emphasizes that knowledge claims are constructed, specifically that they are constructed so as to satisfy the social needs and interests of the knowers. Since different social groups will have different needs and interests, their definitions of knowledge will vary accordingly. Constructionism is thus *relativist, and its proponents reject most claims of *objectivity. Also called "social constructionism."

Constructivism   A theory of learning originally developed by Piaget. His research indicated that children learn by constructing their knowledge on the basis of their experiences. More *subjective versions of constructivism have much in common with *constructionism, though the latter tends to stress social more than individual influences on what counts as knowledge. Some researchers claim that the insights of constructivism should be applied to research methods.

Consumer Price Index   A measure of the average change over time of a fixed group of goods and services. It is used to gauge inflation. In the United States, data for this index are collected monthly or bimonthly from more than 50,000 households by the Bureau of Labor Statistics. See *base year.

Consumption Function   The relation between consumption and other variables—such as income, wealth, and interest rates—expressed as a formula (function). Generally, consumption is a function of (increases with, is caused by) income. See *production function.

Contamination   Refers to research situations in which data or variables that should be kept separate come into contact, often when the *independent and *dependent variables measure the same or similar things. See *confounded.

Dirty test tubes in the chemistry lab are the classic example. A more common example in the social and behavioral sciences could occur when researchers know subjects' scores on an *independent variable before they measure them on the *dependent variable; that knowledge could influence (contaminate) the second measurement. See *double-blind procedure.

Content Analysis   Any of several research techniques used to describe and systematically analyze the content of written, spoken, or pictorial communication—such as books, newspapers, television programs, or interview transcripts. The techniques are often, though not necessarily, quantitative in orientation. Qualitative analyses of content go by names such as *discourse analysis and *narrative analysis.

For example, in a series of interviews you could ask people open-ended questions about different ethnic groups. Later, the audiotapes of these interviews could be transcribed (perhaps entered into a computer program) so that the number of positive and negative adjectives used by interviewees when talking about various ethnic groups could be counted.

A famous early use of content analysis dates from the 1960s. Several psychologists and statisticians used word-frequency analysis to identify previously unknown authors of some of the *Federalist Papers.*

Content-Referenced Test   Another term for *criterion-referenced test.

Content Validity   (a) A measure has content validity when its items accurately represent the thing (the "universe") being measured. Content validity is not a statistical property; it is a matter of expert judgment. Compare *construct, *concurrent, and *convergent validity. (b) In *factor analysis, the ability of a group of measured variables to estimate a *latent variable.

It is easier to give clear examples of *in*validity than validity (definition a). For instance, a test of American history that contained only questions about Civil War battles would not be content valid; its questions would not be representative of the entire subject.

Contextual Effects   The impact on individuals of operating in certain contexts. These are usually studied with *multilevel models or *hierarchical linear modeling. Compare *cohort effects, *conditional effects.

For example, attending a high school in which most of the other students plan to go to college (one context) might influence a student differently than attending a high school in which very few students planned to go to college (another context). Otherwise similar students, attending different types of high schools, might have different propensities to attend college. If so, that would be an example of contextual effects.

Contingency   A relation between variables such that one determines or depends upon (is contingent upon) another. See *conditional probability.

Contingency Coefficient   Short for *Pearson's contingency coefficient, which is a measure of *association for *categorical variables, usually as displayed in a *contingency table. Compare *Cramer's *V*, a more versatile and accurate measure for categorical variables.

Contingency Effect   Another term for *conditional effect or *interaction effect.

Contingency Table   A table of frequencies classified according to two or more sets of values of *categorical variables. Also called a *cross tabulation. It is called a contingency table because what you find in the rows (the usual place for the *dependent variable) is contingent upon what you find in the columns (the usual place for the *independent variable). See *categorical data analysis.

For example, Table C.10 shows the results of a survey of 700 individuals concerning their religious affiliations and attitudes about legal abortion. Table C.11 adds the gender of the respondents; it is an example of a multivariate (more than two variables) contingency table.

Table C.10   Contingency Table (Bivariate): Religious Affiliation and Attitude Toward Legal Abortion

	Religion			
Favor Legal Abortion	*Catholic*	*Protestant*	*Other*	**Total**
Yes	63	278	73	414
No	140	118	28	286
Total	203	396	101	700

Table C.11 Contingency Table (Multivariate): Attitude Toward Legal Abortion by Gender and Religious Affiliation

| | Gender | | | | | | | |
| | Women | | | | Men | | | |
Favor Legal Abortion	Catholic	Protestant	Other	Total	Catholic	Protestant	Other	Total
Yes	40	140	51	231	23	138	22	183
No	64	58	23	145	76	60	5	141
Total	104	198	74	376	99	198	27	324

Continuity Correction   See *Yates's correction for continuity.

Continuous Variable   A variable that can be expressed by a large (often infinite) number of measures. Loosely, a variable that can be measured on an *interval or a *ratio scale. While all continuous variables are interval or ratio, all interval or ratio scales are not continuous, in the strict sense of the term. Compare *categorical variable.

Deciding whether to treat data as continuous can have important consequences for choosing statistical techniques. Ordinal data are often treated as continuous when there are many ranks in the data, but as categorical when there are few. See *discrete variable for further discussion.

For example, height and grade point average are continuous variables. People's heights could be 69.38 inches, 69.39 inches, and so on; GPAs could be 3.17, 3.18, and so on. In fact, since values always have to be rounded, theoretically continuous variables are measured as discrete variables. There is an infinite number of values between 69.38 and 69.39 inches, but the limits of our ability to measure or the limits of our interest in precision lead us to round off continuous values.

GPA is a good example of the difficulty of making these distinctions. It is routinely treated as a continuous variable, but it is constructed out of a rank order scale (A, B, C, etc.). Numbers are assigned to those ranks, which are then treated as though they were an interval scale.

Contrast Coding   A technique for coding *categorical variables that allows researchers to aggregate categories. Unlike dummy coding, which uses a series of 1s and 0s for a multiple category variable, contrast coding might use −1, 0, and +1 or other codes as long as they sum to zero. Also called *orthogonal coding. Compare *effect coding.

Control   To eliminate the effect of. This can be done either by *random assignment in an *experiment or by statistical simulation in nonexperimental research. See *hold constant and *net effects.

C

Control Card   A series of instructions for a computer program telling it what operations to perform on a particular set of data. So called because these instructions (and the data) were at one time entered on computer punch cards. The cards have largely disappeared, but the term lingers on among some researchers.

Control for   Any one of several ways of statistically subtracting the effects of a variable (a *control variable) to see what a relationship would be without it. See *hold constant.

For example, to compare the average incomes of various ethnic groups, we might wish to control for education level. In that way we could measure the effects of ethnic group membership apart from differences in the educational levels among the groups. This would be important, for example, if we were showing how much of the difference in income persisted even when people from different groups had the same education level. See *analysis of covariance and, for an example, *crossbreaks, Table C.15.

Control Group   In experimental research, a group that, for the sake of comparison, does not receive the *treatment the experimenter is interested in studying. Compare *experimental group.

For example, psychologists studying the effects of television violence on attitudes might give subjects a questionnaire to measure their attitudes, divide the group into two, show a videotape of a violent program to one half (the experimental group), and show a nonviolent program to the other half (the control group). A second attitude questionnaire would then be given to the two groups to see whether the programs affected their scores.

Control on   Another way of saying *control for.

Controlled Trial   A *clinical trial in which a new treatment is compared with another treatment (rather than no treatment), usually the standard treatment against which the new treatment is being compared.

Controlled Variable   A term occasionally used for an *independent variable, so called because independent variables are controlled by experimenters.

Control Variable   An extraneous variable that you do not wish to examine in your study; hence you *control for it. Also called *covariate.

Convenience Sample   A sample of subjects selected for a study not because they are *representative but because it is convenient to use them—as when college professors study their own students. Compare *accidental sample, *bias.

Oddly, using this term sometimes tends to legitimize bad practice. Students of mine occasionally say, "I used a convenience sample to gather the data for my project," as though this hard-to-justify method were a reasonable option among the many types of samples such as *random, *systematic, *stratified, and so on.

Convergent Validity   The overlap between different tests that presumably measure the same *construct. See *concurrent validity, *construct validity.

Converging Evidence   Said of the results of multiple studies that lead to the same conclusion. Compare *meta-analysis, *triangulation.

Conversation Analysis   An area of research that draws from psychology and sociology to investigate the rituals and layers of meaning in verbal interactions.

Cook's Distance   The standardized difference between two sets of fitted values; used to identify *influential observations.

Coordinates   Numbers that can be used to plot points on a graph. See *Cartesian coordinates for an example.

Correction for Attenuation   Adjusting a *correlation to estimate what it would have been had the correlated variables been measured without error. See *attenuation.

Correlated Groups Design   A *research design in which some of the *variance in the *dependent variable is caused by a correlation between groups of subjects—or among sets of their scores. Compare *independent samples (or groups). Different tests of statistical significance are required when the groups are correlated rather than independent.

A common form of this research design is a before-and-after study. For example, 5th graders are given a vocabulary pretest. Then, half of them receive an experimental vocabulary enrichment program, the other half the regular language curriculum. At the end of a semester, they are given another vocabulary test (a posttest). The dependent variable is the scores on the posttest. A good part of the students' scores on the posttest could be explained by their pretest scores. For example, students with very large vocabularies before the experiment would still have large vocabularies after it was over. Thus, regardless of the treatment they receive, the scores of students on the pre- and posttests will almost certainly be somewhat correlated, probably highly correlated. The correlated groups' t-test corrects for the fact that large mean differences are less likely when the groups are correlated.

Correlated Samples   Two or more samples in which members of the separate samples share a characteristic or relationship with one another—husband and wife pairs, for example. See *correlated groups design. A *t-test for correlated samples, not the test for *independent samples, should be used with such a design.

Correlation   The extent to which two or more things are related ("co-related") to one another. This is usually expressed as a *correlation coefficient.

Sadly, students often misinterpret a warning about correlations found in elementary textbooks—correlation does not equal causation. I have seen students take this warning so literally as to believe that two correlated variables cannot possibly be causally linked under any circumstances. Less erroneous, and even more widespread, is the mistaken view that a correlation between two variables provides no evidence whatsoever about cause. The evidence from correlations is often weak by experimental standards, but it is evidence, often important evidence, that it would be foolish to ignore. Disciplines as diverse as economics and epidemiology are heavily based on correlational evidence. Textbook warnings would be more accurate were they to read: Correlation does not *necessarily* indicate causation. See *necessary condition.

Correlational Research Design   A design in which the variables are not manipulated. Rather, the researcher uses measures of *association to study their relations. The term is usually used in contrast with *experimental research.

Correlation Cluster   A group of variables that correlate with one another. Compare *factor analysis.

Correlation Coefficient   A number showing the degree to which two *variables are related. Correlation coefficients range from -1.0 to $+1.0$. If there is a perfect *negative correlation (-1.0) between A and B, whenever A is high B is low, and vice versa. If there is a perfect *positive correlation $(+1.0)$ between A and B, whenever one is high or low, so is the other. A correlation coefficient of 0 means that there is no relationship between the variables. (A zero correlation may also occur when two variables are related, but their relationship is not *linear; see *eta.) See also *association, measure of; *correlation matrix; *regression analysis; *scatter plot.

There are numerous ways to compute correlation coefficients, depending on the kinds of variables being studied. Among the most common are the Pearson product-moment, *Spearman's rho, and *Kendall's tau. The term correlation is used by some to refer to any measure of association and by others to refer only to the association of variables measured at an *interval or *ratio level.

Graphically, a correlation is the degree to which two variables form a straight line when plotted on a *scatter diagram. The examples in Figure C. 2 show a strong positive relation (.755) and a weak positive relation (.160). The line through the pattern of points in each scatter diagram is the *regression line, which is the line that comes closer on average to the points than any other possible line.

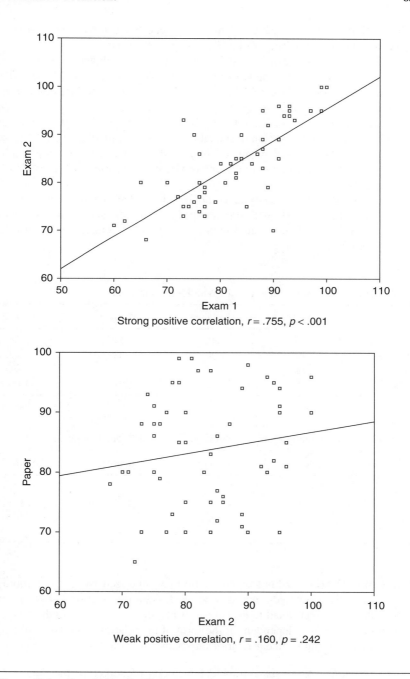

Figure C.2   Correlation Coefficient

C

Correlation Matrix   A table of *correlation coefficients that shows all pairs of correlations of a set of variables. Compare *covariance matrix.

In the following example, correlations between subjects' age, income, and scores on two attitude scales are shown. Note in Table C.12 the series of diagonal correlations of 1.00. These figures reflect the fact that a variable always correlates perfectly with itself. These correlations are often omitted, as they are in Table C.13, since they are self-evident. The upper right (or lower left) portion of the table is usually omitted as well, because it just repeats (as a sort of mirror image) what is printed in the lower left. In short, most correlation matrices are simplified to look like Table C.13, not Table C.12.

Studying either table, you could conclude that the two scales were highly correlated (.91) with one another and that the scores on the scales were more strongly correlated with age (.74 and .68) than they were with income (.33 and .42).

Table C.12   Correlation Matrix (Full; With Redundant Data)

	Age	Income	Scale 1	Scale 2
Age	1.00	.49	.74	.68
Income	.49	1.00	.33	.42
Scale 1	.74	.33	1.00	.91
Scale 2	.68	.42	.91	1.00

Table C.13   Correlation Matrix (Abbreviated; Without Redundant Data)

	Age	Income	Scale 1	Scale 2
Age				
Income	.49			
Scale 1	.74	.33		
Scale 2	.68	.42	.91	

Correlation Ratio   A kind of correlation—symbolized by and commonly known as *eta squared—that can be used when the relation between two variables is not assumed to be *linear. It is a measure of *strength of association, which is independent of the form of the relation—unlike Pearson's *r*, which only shows linear relationship between variables.

Correspondence Analysis   Graphical methods of depicting the degree of association in *cross tabulations.

Counterbalancing In a *within-subjects *factorial experiment, presenting *conditions (*treatments) in all possible orders to avoid *order effects. See *Latin square.

For example, an experimenter might wish to study the effects of three kinds of lighting (A, B, and C) on performance of a visual skill. Subjects could first be placed in Condition A and be given a test of the skill; then they could be put in Condition B and get a second test, and so on. By Condition C and the third test, subjects' scores might go up simply because they had the practice of the first two tests. Or their scores might go down because they became fatigued.

The effects of practice and fatigue could be counterbalanced by rotating the lighting conditions so that subjects would experience them in all possible orders. Since there are six possible orders (ABC, ACB, BAC, BCA, CAB, and CBA), subjects could be divided into six groups, one for each possible order.

Counter Example An example that disproves a general statement, as when the statement "all women hate to study statistics" is countered by the example of Mary, a woman who loves to study statistics.

Counterfactual (Conditional) A statement of what "would have" happened had something occurred that did not in fact occur.

For example, the statement "had the New York stock market crash of 1929 not occurred, the worldwide depression of the 1930s would have been over by 1932" is a counterfactual conditional. Counterfactuals play an important role in theorizing about *cause.

Covariance A measure of the joint or (co-) *variance of two or more variables. See *covariation, *analysis of covariance. A covariance is an unstandardized *correlation coefficient r and it is more often reported as an r. However, the covariance is very important in calculating many multivariate statistics. See *covariance matrix.

For example, suppose we want to see if there is a relation between knowledge of politics (variable X) and political tolerance (variable Y). Our tests of these two variables are each measured on a scale of 1–20. We give the two tests to a *sample of 10 people. The scores and the calculation of the covariance are shown in the following table. Column 1 assigns a number to each individual taking the two tests. Columns 2 and 4 are their results on Test X and Test Y. Columns 3 and 5 subtract the mean of each variable from each individual's score (to get the *deviation scores). Column 6 shows the *product of multiplying column 3 times column 5 (the *cross product). Total Column 6 and divide by the number of cases minus 1 ($10 - 1 = 9$) to get the covariance of X and Y (Cov_{xy}), which equals 11.4. The *correlation r is a standardized version of the covariance, which for these data equals .79.

Table C.14   Covariance (Example of How to Compute)

Column 1 Case	Column 2 X	Column 3 $X - \bar{X}$	Column 4 Y	Column 5 $Y - \bar{Y}$	Column 6 $(X - \bar{X})(Y - \bar{Y})$
01	18	5	16	4	20
02	9	−4	10	−2	8
03	12	−1	11	−1	1
04	17	4	14	2	8
05	13	0	13	1	0
06	8	−5	13	1	−5
07	17	4	16	4	16
08	14	1	11	−1	−1
09	16	3	13	0	0
10	6	−7	4	−8	56
Total	130	0	120	0	103
Mean	13		12		11.4 (covariance)

Covariance Analysis   See *analysis of covariance.

Covariance Components Models   See *hierarchical linear models.

Covariance Matrix   A square matrix formed of the covariances of variables. This is directly comparable to the *correlation matrix, since correlations are standardized (expressed in *z scores) covariances. For both matrices, the numbers on the diagonal are variances. For correlations, the variance is always 1.0. Although the nonstandardized numbers, especially the variances, of the covariance matrix can be very large and hard to manipulate, the covariance matrix is more useful for *structural equation modeling. Also called *variance-covariance matrix.

Covariance Structure Models   See *structural equation modeling, *analysis of covariance structures, *LISREL models, *confirmatory factor analysis.

Covariate   (a) Another term for *independent variable. (b) A variable other than the *independent (or *predictor) variable that correlates with the *dependent (or *outcome) variable. Typically the researcher seeks to *control for (statistically subtract the effects of) the covariate by using such techniques as *multiple regression analysis (MRA) or *analysis of covariance (ANCOVA). Also called *concomitant variable.

Covariation   (a) A state that exists when two things—such as the price and the sales of a commodity—vary together. Measures of *association are designed to capture the degree of covariation. (b) The numerator of a *covariance. See *covariance, *correlation.

Cover Story   Untrue accounts of the purposes of research told to the subjects of the research. Such deception is sometimes necessary in psychological

C

research if subjects' knowledge of the true purposes of the research might *confound the results. See *debriefing and *dehoaxing.

Cov$_{XY}$ *Covariance of X and Y.

CPI   *Consumer Price Index.

CPS   *Current Population Survey.

Cramer's *V*   A measure of *association for *categorical variables based on the *chi-squared statistic. It ranges from 0 to 1 and allows meaningful comparisons of chi-squared values from tables with different sample sizes and different numbers of *cells. Compare *phi coefficient and *Pearson's contingency coefficient, which are also based on chi-squared values, but are applicable to a more limited range of problems and cannot attain a maximum value of 1.0.

Criterion   (a) Short for *criterion variable, that is, the *outcome or *dependent variable. (b) In testing, the specific content or standard against which examinees' performances are compared.

Criterion Group   A group used to *validate a test because its characteristics are known.
   For example, if we wanted to validate a screening test for prospective locksmiths, we could give the test to master locksmiths to see if they performed well on it. If they did not perform well, the test probably would not be a valid measure of skills needed to be a good locksmith.

Criterion-Referenced Test   A test that examines a specific skill (the criterion) that students are expected to have learned, or a level (the criterion) students are expected to have attained. Unlike a *norm-referenced test, it measures absolute levels of achievement; students' scores are not dependent upon comparisons with the performance of other students. Also called a "content-referenced test."

Criterion-Related Validity   The ability of a test to make accurate predictions. The name comes from the fact that the test's validity is measured by how well it predicts an outside criterion. Also called "predictive validity." See also *concurrent validity, which is often held to be another aspect of criterion-related validity.
   For example, the extent to which students' SAT scores predict their college grades is an indication of the SAT's criterion-related validity.

Criterion Scaling   A method of reducing the number of categories of *categorical and *ordinal variables in a *multiple regression analysis. The goal is to make the analysis more manageable by reducing the number of coded *vectors. The technique gets its name from the fact that it involves using the *mean of each group on the criterion (dependent) variable. It is often used in *repeated-measures designs.

C

Criterion Variable   Another term for *dependent or *outcome variable. The term is usually used for nonexperimental studies. In such usage, the *independent variable is called the *predictor or *explanatory variable.

Critical Ratio   The formula that gives the values that define the *critical region.

Critical Region   The area in a *sampling distribution representing values that are "critical" to a particular study. They are critical because when a *sample statistic falls in that region, the researcher can reject the *null hypothesis. (For this reason, the critical region is also called the "region of rejection.")

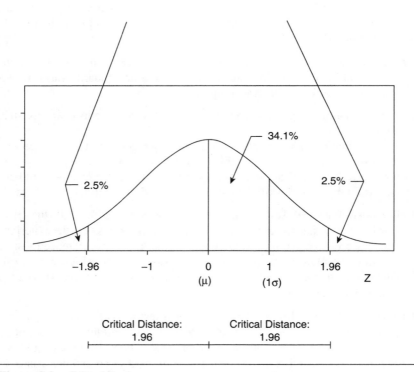

Figure C.3   Critical Region

Critical Theory   A term applied to several approaches to research and scholarship, most of which blend *relativism with left-wing political commitment. Often associated with the work of Jürgen Habermas (1929–), critical theory stresses the distinction between the natural and the social sciences and thus rejects *positivism. Variants include critical ethnography, critical pragmatism, critical race theory, and so on. See *Geisteswissenschaften.*

Critical Values   (a) The values that determine the *critical regions in a *sampling distribution. The critical values separate the obtained values that will and that will not result in rejecting the *null hypothesis. See illustration under *critical region.

(b) Tables of values for *test statistics, which when exceeded, enable the researcher to reject the null hypothesis. Such tables are used to interpret the results of *t-tests, *chi-squared statistics, or *F ratios. The values on such tables are in the *metric of these tests; they are not expressed in *raw data.

For example, when a normal distribution with an *alpha level of .05 is used as a test of significance, the critical value is 1.96. If the *test statistic is greater than 1.96, then the null hypothesis is rejected.

Cronbach's Alpha   A measure of internal *reliability or consistency of the items in an *instrument or *index. It is a widely used form of *Kuder-Richardson formula 20 (KR20); but, unlike KR20, it can be used for test items that have more than two answers, such as *Likert scales. Cronbach's alpha is a measure of the *intercorrelation of the items and estimates the proportion of the variance in all the items that is accounted for by a common *factor. Like other *reliability coefficients, it ranges from 0 to 1.0. Scores toward the high end of that range (e.g., above .70) suggest that the items in an index are measuring the same thing. Also called "alpha coefficient" and "coefficient alpha."

Cronbach's alpha is replacing KR20 because it is more versatile. An alpha on dichotomously scored items gives the same reliability score as KR20.

Crossbreaks   Also called cross tabulations ("tabs") and cross partitions. A way of arranging data about categorical variables in a matrix so that relations can be more clearly seen. This is not to be confused with a *factorial table, in which two or more variables are related to a third. While not all researchers make these distinctions in the terms, the concepts are quite distinct. Compare *contingency table.

For example, Table C.15 is a crossbreaks table. It shows the relations among race, sex, and graduation rates. Table C.16, on the other hand, is a factorial table where the influence of two variables (sex and education) on a third (median annual income) is shown.

Table C.15   Crossbreaks Table Showing the Percent of High School and College Graduates Among Persons Aged 25–29, by Race and Sex (2002)

	High School or More	Bachelor's Degree or Higher
Men	84.7	26.9
Women	88.1	31.8
Whites	85.9	29.7
Blacks	86.6	17.5

Table C.16   Crossbreaks Table (Compared to Factorial Table) Showing Median
                    Annual Income of Full-Time Workers 25+ Years Old, by Sex and
                    Education Level (2000)

	High School or More	Bachelor's Degree or Higher
Men	$34,303	$61,868
Women	$24,970	$42,706

Cross-Case Analysis   Another term for *comparative method.

Cross-Cultural Method   See *comparative method.

Crossed Factor Design   The usual way two or more factors are combined in
a *factorial design. When every level of one factor appears within every
level of the other factor(s), they are said to be (completely) crossed. The
opposite of crossed is "nested." See *nested design for illustrations of the
two designs.

Cross-Lagged Models   *Regression models for longitudinal or *panel data.
The technique is designed to determine whether *independent variables at
one session or *wave are related to *dependent variables in the next wave
of data gathering.

Cross-Level Inferences   Making inferences about one *level of analysis
based on data about another, such as making inferences about individuals
based on data about groups. See *ecological fallacy.

Crossover Design   A type of *longitudinal, *within-subjects study in which
subjects receive different treatments at different times. Treatments are
allocated randomly.

Crossover Interaction   Another term for *disordinal interaction.

Cross Partition   A combination of two or more *partitions.
        Say, for example, we were studying unemployment rates. We could look
at them in general (for all people) or we could partition the data by group.
We could examine unemployment among men and women (one partition)
or among different ethnic groups (a second partition). Cross partitions
would combine the first two partitions (gender by ethnicity) so that we
could study groups such as white women, Hispanic men, and so on.

Cross Products   Short for cross-products deviation scores. A step in the
calculations to determine the *covariance; the cross products are obtained
by multiplying the *deviation scores of one variable times those of another.
See *covariance (Column 6 of Table C.14) for an example.

Cross-Products Ratio   Another term for *odds ratio.

Cross-Sectional Data Data gathered at one time. Compare *time-series data, *panel study.

Cross-Sectional Study A study conducted at a single point in time by, so to speak, taking a "slice" (a cross section) of a population at a particular time. Compare *panel study, *longitudinal study, *cohort-sequential design.

Cross-sectional studies provide only indirect evidence about the effects of time and must be used with great caution when drawing conclusions about change. For example, if a cross-sectional survey shows that respondents aged 60–65 are more likely to be racially prejudiced than respondents aged 20–25, this does not necessarily mean that as the younger group ages it will become more prejudiced—nor does it necessarily mean that the older group was once less prejudiced.

Cross Tabulation A way of presenting data about two variables in a table so that their relations are more obvious. Also called *contingency table and *crossbreaks table (see that entry for an example). Compare *factorial table.

Cross Validation Using one *sample or one part of a sample to develop a theory and select appropriate statistics and then using another sample or part of a sample to test (validate) the theory and the performance of the statistic. Cross validation is important, because using the same data to develop a theory and to test it is a form of circular reasoning or *tautology.

Crucial Experiment An experiment or other study that decisively tests a theory or hypothesis. There is some controversy about whether any one experiment can be crucial in this sense, particularly in the social and behavioral sciences.

CRV Coefficient of *Relative Variation.

Cultural Capital Cultural resources (such as verbal fluency and educational credentials) that one can use to obtain income or other resources. Compare *capital, *social capital, *human capital.

Cultural Relativism The belief that human thought and action can be judged only from the perspective of the culture out of which it has grown.

For example, a person who is generally opposed to male chauvinism, but who is also a cultural relativist, might conclude that one should not condemn male chauvinism if it could be seen as an integral part of the culture of a particular ethnic group. Of course, this relativistic judgment could itself be relative to another cultural group—middle-class Western intellectuals, perhaps.

Cumulative Frequency For any value or *class interval in a *frequency distribution, the total up to and including that value or interval.

C

For example, the grades of 43 students on an examination are shown in the following table. The cumulative frequency for the class interval 70–79 is 13, which is the total up to and including that interval $(4 + 9 = 13)$.

Table C.17   Cumulative Frequency: Final Examination Grades of 43 Students

Interval	Frequency	Cumulative Frequency
A 90–99	12	43 $(4 + 9 + 18 + 12)$
B 80–89	18	31 $(4 + 9 + 18)$
C 70–79	9	13 $(4 + 9)$
D 60–69	4	4

Cumulative Frequency Polygon   A graphic representation of a *cumulative frequency distribution. In the following example, a more detailed version of the data from Table C.17 is graphed. The *cumulative standard normal distribution is another example.

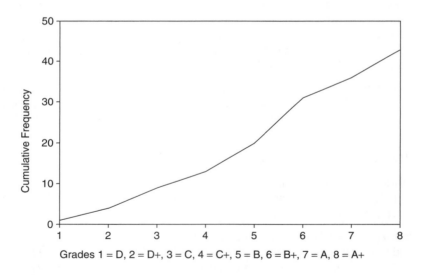

Figure C.4   Cumulative Frequency Polygon

Cumulative Standard Normal Distribution (CSND)   If you remember the normal distribution and the corresponding percentile scores, the concept of the CSND is easy to understand. A score that is 2 standard deviations below the mean $(z = -2.0)$ would fall in the 2nd percentile and would have a *cumulative frequency or probability of 2%. A score 2 standard deviations above the mean ($z = +2.0$) would be in the 98th percentile and have a cumulative probability of 98%, and so on. The left side of the CSND looks like any other standard normal distribution, but the curve reverses direction at the midpoint. The CSND is used in *probit regression. The following graphic shows an empirical distribution that closely approximates a CSND.

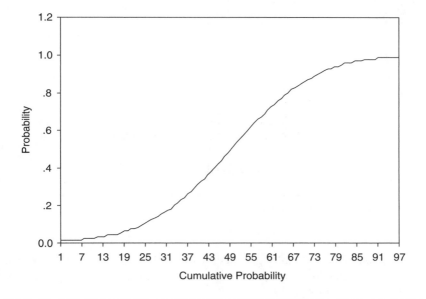

Figure C.5   Cumulative Standard Normal Distribution

Cumulative Scale   A *scale, strictly speaking. See *Guttman scale, *index.

Cup   In *set theory, the symbol ∪ , meaning "or." It is used to indicate the *union of two sets. Compare *cap.

Current Population Survey   An annual survey conducted by the U.S. Census Bureau. About 60,000 households are sampled and studied, mainly regarding income and employment status.

Curvilinear Regression   Another term for *polynomial regression. See *spline regression.

Curvilinear Relation (or Correlation)   A relationship between two *variables that, when plotted on a graph, forms a curve rather than a straight line (a *linear relationship). See *eta, *polynomial regression analysis.

   For example, the relation between physical strength and age is curvilinear. As children get older they tend to get stronger, and this continues into

C

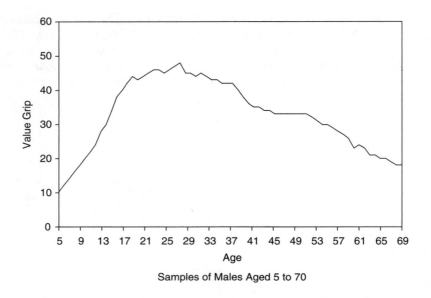

Figure C.6   Curvilinear Relation

---

adulthood; but as adults get older they tend to get physically weaker. Figure C.6 graphs the relation between age and grip strength for a sample of males ages 5 through 70.

CV   (a) *Critical value. (b) Coefficient of *Variation.

Cybernetics   A discipline specializing in the study of communications systems, particularly as they relate to control mechanisms, as when computers run robotic assembly lines. Compare *artificial intelligence, *information theory.

Cycle   Any regular variation (down and back up or up and back down) in *time-series data, usually after *seasonal adjustments have been made.

Cyclic Data   *Time-series data arranged on a measurement scale in which the numbers or categories recur, as 12 o'clock is followed by 1 o'clock, December 31 by January 1, and so on. Special statistical techniques are required to describe and analyze most types of cyclic data.

D (a) Used to symbolize a wide variety of *difference scores, such as the difference between ranks when computing *Spearman's rho. (b) Abbreviation for *deviance. (c) Abbreviation for a standardized *effect size index, also called Cohen's *d*. It reports the difference between the *means of two groups in terms of their common *standard deviation. When *d* = 1.00, for example, the mean of one group is one standard deviation above that of the other group. A lowercase *d* is used for the *parametric statistic; an uppercase *D* is used for the *nonparametric statistic. See *Somers's d, *Kim's d, *D test.

***D*²** Symbol for the *Mahalanobis distance statistic, which is a widely used test for multivariate *outliers.

Dandekar's Correction A method of adjusting the calculation of a *chi-squared statistic for a *two-by-two table. Compare *Yates's correction.

Data Information collected by a researcher; any organized information. ("Data" is the plural term; the singular is "datum," but usage varies.) Data are often thought of as statistical or quantitative, but they may take many other forms as well—such as transcripts of interviews or videotapes of social interactions. Nonquantitative data such as transcripts or videotapes are often *coded or translated into numbers to make them easier to analyze.

Database (or Data Base) (a) A collection of data organized for rapid search and retrieval, usually by a computer; often a consolidation of many records previously stored separately. (b) Sometimes used loosely to mean *sample size.

Data Curve A line formed by connecting the *data points on a graph. For examples see *frequency polygon and *learning curve.

D

Data Entry   The process of preparing data for use by a computer, or of putting data into a computer, usually by using a keyboard.

Data File   A collection of *data records organized for retrieval and analysis.

Data Filter   Criteria used to select data for inclusion in a data file that will be analyzed.

For example, in a study of average incomes of full-time workers to be drawn from a large database, you might use the status "unemployed" or "age less than 16" as a filter to reduce the size of your data file. Unemployed people, by definition, or people under 16 years of age, by law, cannot be employed full time.

Data Mining   "Digging around" in a large *database in order to discover relationships among variables or, sometimes, until you find a statistical association that "demonstrates" something you would like to demonstrate. With the increasing availability of huge databases and powerful computer programs, the importance of data mining has grown significantly. Compare *fishing expedition. Contrast *hypothesis testing.

Data Matrix   A grid for storing and subsequently locating data, usually in a computer format.

For example, suppose you surveyed five people and asked each of them four questions. The results of the survey could be put in the following type of data matrix. The rows (or *records) represent the persons interviewed (the *respondents, *cases) and the columns (or *variables) their answers to the questions (the *variables). In the example, the "a" shows the location (the *cell) of respondent number 01's answer to question number 1 (Q1), "f" shows where number 2's answer to Q2 would be placed, "t" indicates where number 5's answer to Q4 would be found, and so on.

Table D.1   Data Matrix of Hypothetical Survey

	Question			
Respondent	*Q1*	*Q2*	*Q3*	*Q4*
01	a	b	c	d
02	e	f	g	h
03	i	j	k	l
04	m	n	o	p
05	q	r	s	t

Data Point   An individual piece of data; a datum. Often, the point at which two values intersect on a graph, as in the following example, where the data point for a subject who is 66 inches tall and weighs 150 pounds is circled.

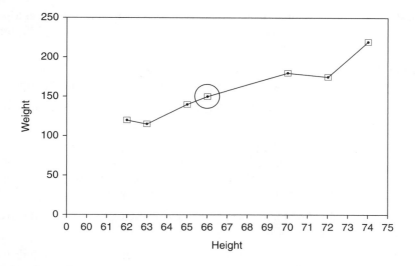

Figure D.1   Data Point

Data Reconstruction   In *meta-analysis, any of various methods for using research findings to calculate data not reported. This is done so that the person conducting a meta-analysis can compare or combine the results of different studies that do not report their results in the same way.

One common form of data reconstruction occurs when a study reports the total number of subjects and the percentage of them that fall into various categories, e.g., "of the 1,500 people interviewed, 52% were females, 18.6% of whom had four or more years of college education." A researcher could reconstruct the absolute numbers from these percentages: 780 females were interviewed; 145 of them had four or more years of college.

Data Record   A grouping of data composed of one or more lines. There is one record for each *subject or *case in a study. A record is part of a *data file. The columns indicate the location of the data about each variable for each case, as in the following example in Table D.2 giving the age, height, and weight for three subjects. Compare *data matrix.

Table D.2   Data Record

	Fields		
	Variable 1 (Age)	Variable 2 (Height)	Variable 3 (Weight)
Case/Record 1	27	72	173
Case/Record 2	35	70	180
Case/Record 3	25	63	125

Data Reduction   Summarizing large amounts of data, usually by *descriptive statistics such as measures of *central tendency, but also by *factor analysis and *principal components analysis.

Data Set   A collection of related *data items, such as the answers given by respondents to all the questions on a survey. Compare *data file, *database.

Datum   Singular of *data. An individual number, symbol, or other item of information, such as $4.25, 72 inches, or 91%. See *cell.

Debriefing   Explaining the purposes of an experiment to subjects after their participation in it is over. This is required, legally as well as ethically, especially when the experiment has involved deceiving subjects or has in any way put them at risk of some harm. See *dehoaxing, *desensitizing.

Decidable   Said of problems that are solvable, particularly with an *effective procedure or an *algorithm.

Decile   One of the nine points that divides a *frequency distribution into 10 equal parts. So, 10% of the cases fall below the first decile, 20% below the second, and so on.

   For example, if there were 120 million wage earners in the United States, a researcher might divide them into ranked tenths (or deciles) of 12 million each, the lowest earning tenth, the second-lowest earning, and so on. This would facilitate comparisons, such as of the average earnings of people in different deciles.

Decile Range   The difference between the ninth and first decile; the middle 80% of a distribution. Compare *interquartile range.

Decision Error   A mistake made when deciding whether or not to reject the *null hypothesis. Compare *Type I error and *Type II error.

Decision Function   A rule of procedure in sampling that specifies whether sufficient data have been collected or whether further observations need to be made. Compare *sequential sampling.

Decision Problem   The problem of figuring out whether a problem is *decidable.

Decision Rule   (a) A statement specifying when a statistic we are about to compute will lead us to reject or not reject the *null hypothesis. (b) Any procedure for making a decision. See *decision function.

   An example of (a) would be a rule such as the following: If the differences between the diabetes mellitus rates for samples of ethnic groups A, B, and C are equal to or greater than 4%, we will reject the null hypothesis of no differences between the groups. Examples of (b) include the criteria used to *operationalize a variable, such as: We will record only rates that increase or decrease, not those that remain stable. Or, to be elected, a candidate need only obtain the most votes; a majority is not required.

Decision Table   A table depicting the alternatives to be considered in a given problem, along with the outcomes of each alternative and action(s) to be taken.

For example, suppose you went to a doctor who told you that you had a terrible degenerative disease. According to this doctor, the *probability is very great (.90) that it will kill you within a year. You can reduce your chances of death from the disease (but only somewhat, to .80) by a radical change of diet. There is an operation, but it is risky: 50% of those who have the operation are cured, but 50% die on the operating table. The following decision table shows your options and the likely outcomes. Without any other information, it looks as though your "best" choice might be to have the operation, but "best" in this case, as in so many others, must ultimately be based on your subjective values, not on arithmetic.

Table D.3   Decision Table

Options	Outcomes, Probabilities of Survival
Do Nothing	.10
Change Diet	.20
Operation	.50

Decision Theory   An interdisciplinary area of research that focuses on how to select good ways of making decisions based on evidence. The basic method is to compare costs and benefits (often called *utilities) and study these along with the *probability of the various alternatives. Decision theory originated in problems of economic decision making, but has become increasingly associated with statistics and hypothesis testing. See *game theory, *minimax principle.

Decision Tree   A graphic representation of the alternatives in a decision-making problem.

For example, suppose you were considering buying some high-risk stock. The cost of the stock is $5,000. If the company in which you are investing is successful, your stock will be worth $40,000. If it fails, you lose the $5,000. On the basis of past performance of such companies, you esti-mate that the probability of success for this one is .10 or 10%. The solution to the problem of whether or not to invest can be summarized in the follow-ing decision tree.

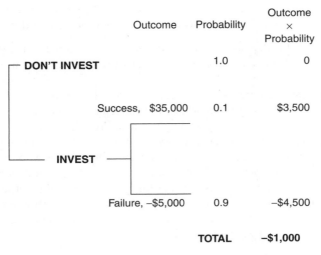

	Outcome	Probability	Outcome × Probability
DON'T INVEST		1.0	0
INVEST	Success, $35,000	0.1	$3,500
	Failure, −$5,000	0.9	−$4,500
		TOTAL	−$1,000

DECISION: Don't invest; over the long run you would lose $1,000 per investment.

Figure D.2   Decision Tree

Decode   To translate or determine the meaning of *data that have been *coded.

Decomposition   (a) Splitting a *time series into its component parts: *trends, regular *fluctuations, and random fluctuations. (b) A similar division of a *correlation coefficient or a *regression coefficient into direct effects, indirect effects, and dependence on common causes. Compare *partitioning of variance. See *effects analysis.

Deconstructionism   An approach ("method" would not be appropriate) to studying texts that rejects the assumption that they have logical meanings and argues for "demystifying" texts instead of deciphering them. Interpretations of texts are little more than word games, according to some proponents of deconstructionism, such as Jacques Derrida (1930–2004).

Deduction   (a) A conclusion that follows logically from known (or assumed) principles, that is, a conclusion arrived at using *deductive methods. (b) The process of reasoning that moves from general principles to conclusions about particular instances. See *deductive. Compare *induction.

Deductive   Said of conclusions derived by reasoning rather than by data gathering; or, research methods using such reasoning. A *hypothesis is often arrived at by deduction from a *theory or other assumed truth; the hypothesis can then be tested using *inductive (data-gathering) research methods. For example:

a. Prejudice is the product of ignorance.
b. Education reduces ignorance.
c. Therefore, the prejudice level will go down as education level goes up.

The assumptions or theories are (a) and (b); (c) is the deduction. It can be turned into a research question or hypothesis. For instance, one could do survey research to see whether education levels and prejudice levels vary *inversely. Are people with low levels of education more likely to give prejudiced answers to survey questions, and vice versa?

Default   In computer jargon, said of a disk or a drive or a value. It is the one the computer *software assumes you mean when you do not tell it otherwise, that is, when you "default" on your obligation to specify what you mean.

For example, if you give your computer the instruction "save this file" but do not specify where to save it, the computer will save it on the default disk or drive. The term is sometimes used more broadly, as in ".05 is the default *alpha level used in this research." This means "unless I say otherwise, it is .05."

Degrees of Freedom   Usually abbreviated df. The number of values free to vary when computing a statistic; the number of pieces of information that can vary independently of one another. The df tells you how much data was used to calculate a particular statistic. This number is necessary to interpret a *chi-squared statistic, an *F ratio, and a t-value. In the t-test, F test, and in regression analyses, the df are determined mostly by the number of observations of cases or subjects. In *z-tests, degrees of freedom are not used.

Many people find the concept of df difficult, but the practical application relatively easy; that is, statistics texts contain clear rules for how to calculate and use the df to interpret a statistic.

The degrees of freedom in a *cross tabulation provide the clearest example. The df are computed by multiplying the number of rows minus 1 times the number of columns minus 1; $df = (R - 1)(C - 1)$. Thus the more categories the variables are broken into, the higher the degrees of freedom.

For example, suppose a professor with 130 students gave a test and tabulated the scores. Table D.4 is a *two-by-two (2×2) table; it has two rows and two columns. Using the formula, $df = (R - 1)(C - 1) = (2 - 1)(2 - 1) = 1 \times 1 = 1$. Table D.4 has 1 df, which means, among other things, that if you know one of the *cell values and the totals (or *marginals), you can figure out the other three. For instance, if 40 men passed, it is easy to figure out how many men failed ($70 - 40 = 30$), how many women passed ($90 - 40 = 50$), and how many women failed ($40 - 30 = 10$).

Table D.5 is a 2×5 table. Using the formula, $df = (R - 1)(C - 1) = (2 - 1)$ $(5 - 1) = 1 \times 4 = 4$. This means that if you know four of the cell values and the marginals, you can compute the other six—because they are no longer free to vary once four are determined.

Table D.4   Degrees of Freedom (2×2 Table): Test Grade by Sex

	Pass	*Fail*	*Total*
Men			70
Women			60
Total	90	40	130

Table D.5   Degrees of Freedom (2×5 Table): Test Grade by Sex

	A	*B*	*C*	*D*	*F*	*Total*
Men						70
Women						60
Total	20	50	20	30	10	130

Dehoaxing   A form of *debriefing of subjects in an experiment after their participation is concluded. When the experimental design requires deceiving ("hoaxing") subjects about themselves, dehoaxing involves convincing them that they have been deceived. The idea is to eliminate any undesirable effects the deception might have had. Compare *desensitizing. Dehoaxing should also be considered after a *participant observation study in which the investigator has not revealed that she or he was doing research.

For example, if a learning experiment involved studying the effects of believing that one is not good at learning a particular subject, the researchers might give all subjects an aptitude test. The *experimental group might be told that they did quite poorly and demonstrated low aptitude, regardless of how they actually performed. The *control group might not be informed one way or another about their scores. Then, both groups could be given the same kind of learning task to see whether the experimental group (those who had been falsely told they had low aptitude) performed any differently from the control group. It is generally considered the researcher's ethical responsibility in such circumstances to dehoax the subjects, to convince them that they were deceived, and that, in this example, they are not in fact low in aptitude. It may sometimes be difficult to convince subjects; the researcher's credibility can be reduced by the fact that he has just admitted lying: "I was lying before, but now I'm going to tell you the truth."

Deliberate Sampling   Another term for *quota sampling. See *purposive sampling.

Delimiting Variables   Variables that specify the nature of a population or a sample. For example, a sample of female college students would have three delimitations: female, student, and college.

Delphi Technique   A method of survey research developed by the RAND Corporation requiring repeated surveying of the same respondents on the same issue or problem so that they can come to an informed consensus. In business, sometimes called "jury of executive opinion."

    For example, managers in a large organization might be sent questionnaires asking them to rank a list of the organization's priorities and to explain their reasoning. Later surveys (a minimum of four) provide each respondent with information about how the others have answered questions on the prior surveys.

Delta [Δ, δ]   A Greek letter most often used to symbolize one form or another of difference. Compare *D. For example, delta-L^2 is the difference between two L^2s. Delta-R^2 indicates the change in R^2 attributable to adding an *independent variable to a *regression equation.

Demand Characteristics   Any of the numerous potential cues available to subjects in experimental research, regarding the nature and purpose of the study, that might influence the subjects' reaction to the experimental treatment.

    For example, an experimenter might, without realizing it, nod encouragingly when subjects act in ways that seem to be supporting the research hypothesis; the experimenter seems to be "demanding" certain behavior from subjects. One way to reduce this problem is to use *double-blind procedures.

Demography   The study of *variables in human populations (such as births, deaths, health, fertility, and migration) and of the variables that cause them to change. Demographic variables are often used by researchers in other disciplines as *background variables.

Dendogram   A kind of *tree diagram representing shared characteristics of cases; similar cases are categorized onto separate branches. It is used in *cluster analysis.

Denominator   Another term for the *divisor in division; the part of a fraction that is below the line.

Density Curve   A graphic representation of a distribution of scores or values that takes the form of a smooth curve. It indicates the proportion of scores in a distribution as the area under the curve. The total area is 1.0. If, for example, 30% of the scores in a distribution fell between 65 and 85, then those scores would be represented by 30% of the area under the curve. *Normal curves are the best known examples, and like all density curves, they are idealized or theoretical rather than exact representations of actual distributions.

Density Function   The equation for a *theoretical relative frequency distribution. Also called "probability function." See *probability distribution, definition (b), and *probability density function.

Dependent Event   In *probability theory, said when the occurrence of one event changes the probability that a second will take place; the probability of the second event is then "dependent" upon the first. See *conditional probability. Compare *independent event.

Dependent Interviewing   In *panel studies in which *respondents are reinterviewed, dependent interview questions are those based on answers to questions from an earlier session (*wave). For example, "Last year you said that when you finish college you were planning to go to graduate school. What are your current plans?"

Dependent Samples   Another term for *correlated samples or groups; said of research groups that are not drawn independently from a population. Dependent samples occur most commonly in before-and-after studies when two measures are taken on the same subjects. Dependent samples require different *test statistics than independent samples.

Dependent Variable   (a) The presumed effect in a study; so called because it "depends" on another variable. (b) The variable whose values are predicted by the *independent variable, whether or not they are caused by it. Also called *outcome, *criterion, and *response variable.

For example, in a study to see if there were a relationship between students' drinking of alcoholic beverages and their grade point averages, the drinking behavior would probably be the presumed cause (independent variable); the grade point average would be the effect (dependent variable). But, it could be the other way around—if, for instance, one wanted to study whether students' grades drive them to drink.

Note: Some authors use the term "dependent variable" only for *experimental research; for *nonexperimental research they might use (or argue that others should use) *criterion variable or *outcome variable. Most commonly, however, dependent variable is used in both experimental and nonexperimental research.

Derived Statistics   Statistics calculated on the basis of other (simple or primary) statistics. Compare *raw data.

D

For example, say you had *data describing the total number of murders last year in all U.S. cities and the total populations of those cities. You could use those primary statistics to compute the murder rates—statistics derived by dividing the number of murders in each city by its population. Other examples of derived statistics include *percentile ranks and *standard scores. Compare *data reconstruction.

Deseasonalizing   Eliminating seasonal variations from *time-series data. Often reported as "seasonally adjusted," as in: The seasonally adjusted unemployment rate for July was at a 5-year low.

Descending Order   An order that begins with the highest value and moves to the lowest. The opposite is *ascending order.

Descriptive Research   Research that describes phenomena as they exist. Descriptive research is usually contrasted with *experimental research, in which environments are controlled and *subjects are given different *treatments. Note that *inferential (not only descriptive) statistics may be used in descriptive research.

Descriptive Statistics   Procedures for summarizing, organizing, graphing, and, in general, describing quantitative information. Often contrasted with *inferential statistics, which are used to make inferences about a *population based on information about a *sample drawn from that population. Descriptive statistics are sometimes also contrasted with analyses of causal relations. And they are also sometimes contrasted with *multivariate statistics, in which relations among three or more variables are examined; in that case "descriptive" is used to mean *bivariate or *univariate.

Desensitizing   A form of *debriefing subjects after their participation in an experiment. The purpose of desensitizing is to enable subjects to cope with any negative information they may have acquired about themselves as a result of an experiment. Compare *dehoaxing, *Milgram experiments.

Design   Short for *research design, that is, the plan a researcher will follow when conducting a study. See *protocol.

Design Effect (or Factor)   The influence of a sampling design, such as *cluster sampling or *stratified sampling, on a study's *standard error. Usually reported as a comparison to what the standard error would have been with *simple random sampling.

Determination, Coefficient of   A statistic that indicates how much of the *variance in one variable is determined or explained by one or more other variables; more strictly, how much the variance in one is associated with variance in the others. It is calculated by squaring the *correlation coefficient. Thus it is abbreviated r^2 in *bivariate analyses and R^2 in *multivariate analyses. Also called "index of determination." See *strength of association.

For example, one might find a statement like the following in a research report: "Education level attained explains 22% of adult occupational status $(r^2 = .22)$."

Determinism The theory that all events and behaviors are determined or caused by prior events, conditions, and the operation of natural laws. Under the assumptions of strict determinism, there are no random events, and people do not have free will. "Soft" versions of determinism exist; they might be thought of as "influence-ism"; these are more common among social and behavioral scientists than the strict variety.

Deterministic Model A *causal model that contains no random or probabilistic elements; one in which all causes and values are known and all the *variance in the *dependent variable(s) can be explained. Compare *stochastic model.

Detrending To *control for trends in one's data caused by *variables in which one is not interested.

For example, in a study of the effects of education level on productivity level, you might find that the trend toward increasing education was associated with a trend toward increasing productivity. But productivity can also go up for reasons unrelated to education, such as increased capital investment. By controlling any trends toward increased capital investment, you could then focus better on the relation of education to productivity.

Detrend Normal Plot A scatter plot used to check visually the assumption that one's sample is drawn from a normally distributed population. If the sample is from a normal population, the points should form no pattern. Also called "detrend probability plot." See *normal probability plot for graphic illustrations; see also *Shapiro-Wilks test.

Deviance A measure of the degree to which a *model explains the *variation in a set of data when using *maximum likelihood estimation. It involves comparing the *saturated and *unsaturated models. Abbreviated D.

Deviate Another term for *deviation score.

Deviation Score A statistic indicating how much the *mean score of a group of scores differs (deviates) from an individual score. It is obtained by subtracting the group mean from the individual score. Also known as an *error score. See *covariance for an example.

df *Degrees of freedom.

Diachronic Said of research that studies events as they occur or change over time. Often contrasted with *synchronic. Compare *panel study, *event history analysis.

D

Diagnostic Test   A procedure in medicine, psychology, or education designed to determine whether a subject has a particular condition, or the extent of that condition. See *sensitivity and *specificity.

Diagraph   Short for *directed graph.

Dialectic   A method of reasoning that proceeds by developing contradictions to propositions and then discovering ways to resolve those contradictions so as to discover new ideas and advance thought. Although employed by many philosophers since ancient Greece, the dialectic is perhaps best known as it was used by Karl Marx (1818–1883), who held that history progresses dialectically, through the conflict of opposing classes.

Dichotomized Variable   A continuous variable that has been divided into two categories. For example, one might use income data to create two categories, poor and not poor. Compare *dichotomous variable, in which the two categories occur naturally, that is, they are not created by the researcher. The distinction between dichotom*ized* and dichotom*ous* variables is important when selecting appropriate measures of *association.

Dichotomous Variable   A *categorical variable that can place subjects or cases into only two groups, such as male/female, alive/dead, or pass/fail. Compare *dichotomized variable.

DIF   *Differential item functioning.

Difference of Proportions   A method for comparing proportions for *dichotomous variables. One proportion is subtracted from the other. The result ranges from −1.0 to +1.0, with zero indicating that the two variables have identical conditional probabilities on a dependent variable.

   For example, a study of a medical treatment that resulted in Cure or No Cure could use difference of proportions to describe results for men and women. If the proportion of women treated who were cured was .60 and the proportion of men was .45, then the difference of proportions would be .15.

Difference Scores   Measurements obtained by subtracting pretest (before) from posttest (after) scores. Sometimes called "gain scores," even when the goal is to lower the score, for example, to reduce the time required to complete a task or to lose weight. Also called "change scores."

   Difference scores are often *standardized by transforming pre- and posttest scores into *standard scores before subtracting. The size of the difference is often adjusted by *controlling for *nuisance variables.

Difference Sign Test   A test of *time-series data to see if a *linear trend exists. It is calculated by counting the number of times the series increased—for example, the number of times the Dow Jones average went up over the past year. See *sign test.

D

Differential Item Functioning (DIF)   In *item response theory (IRT), differential item functioning occurs when test items function differently for different groups, such as ethnic and *SES groups, that is, when different groups have different probabilities of giving correct responses to items. IRT can be used to find items for which such differences do not exist.

Diffuse Comparisons   Techniques used in *meta-analysis to compare the amount of heterogeneity in the analyzed studies' *significance levels and *effect sizes. The more heterogeneity or "diffuseness," the harder it is to integrate the studies. See *focused comparisons. Compare *divergent validity.

Diffusion of Treatments   A threat to the *validity of a study arising from communication among the subjects, in particular, when the communication results in the experimental *treatment being spread ("diffused") among *control group subjects. Also called "diffusion effect." Contrast with *double-blind procedure.

For example, if a new technique were being tested in a chemistry laboratory to see if it led to quicker and more accurate analyses, it could be tried out by a sample of the lab workers. They would be the *experimental group. The rest of the workers would be the *control group. To measure the effectiveness of the new technique, the productivity and efficiency of the two groups of workers would be compared. But, if the experimental group liked the new technique and told their friends in the control group about it, and they also started using it, the validity of any comparison between the two groups would be doubtful at best—because of the diffusion effect.

Digital Data   (a) Information represented by numbers (digits), such as time on a digital watch; often contrasted with *analog data, such as time represented by movement of a watch's hands. (b) Loosely used to mean *binary, as in "digital computer," that is, a computer that uses information represented in the form of 1s and 0s. The 1s and 0s represent electronic computer switches that are "on" (1) or "off" (0).

Dimension   Clusters of related *indicators. For example, for the dimension public health, the indicators could be the infant mortality rate, average life expectancy, a measure of access to medical treatment by citizens, and so on.

Dimensionality   The number of aspects or "dimensions" a *construct has. Is "tolerance," for example, one attitude or a cluster of related attitudes? If it is one attitude, it is said to be a unidimensional construct; if more than one, it is a multidimensional construct.

***D* Index**   See *D.

Direct Correlation   Another term for *positive correlation, that is, a correlation in which the values of the variables tend to move in the same direction. Compare *inverse correlation.

Directed Graph   A pictorial representation of a *dominance matrix used in *game theory and similar problems.

The following example shows who won in a chess tournament among four players. An arrow indicates the winner. Anne, for example, won her match with Carl. See *dominance matrix for a table showing the same data.

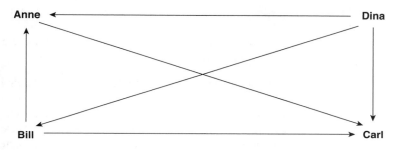

Figure D.3   Directed Graph

Directional Hypothesis (or Test)   An *alternative hypothesis that specifies the direction of difference (greater or smaller) from a *null hypothesis. Sometimes called a *one-tailed test. Directional tests are easier to pass than are nondirectional. When in doubt about which to use, most experts recommend the nondirectional.

For example, say the null hypothesis were: There is no difference in manual dexterity between Group A and Group B. An example of a directional alternative hypothesis would be: Group A's average dexterity quotient is greater than B's.

Directionality Problem   Uncertainty about the direction of causal relations among variables, especially in *correlational research designs. (The difference between "directionality" and "direction"? Three syllables.)

For example, if aggressive people watch lots of violence on TV, it could be that TV causes them to be aggressive, or that aggressive people choose to watch violent programs, or both. Or neither, but that is a different issue. See *spurious correlation.

Direct Relationship (or Correlation)   A relation between two *variables such that when the value of one goes up or down, so does the other. Also called *positive relationship. Compare *inverse relationship.

For example, hours spent studying and grade on an examination might be directly related, that is, the more hours students study the better they do; the fewer hours they study, the worse they do.

Disaggregate   To separate, for purposes of analysis, the parts of an *aggregate statistic. Compare *decomposition.

D

For example, if we were interested in trends in average SAT scores over the past 20 years, we might want to disaggregate the data so that we could look at separate trends for males, females, Blacks, Whites, students in and not in college preparatory programs, and so on. Or we could disaggregate by studying different dimensions of the test.

Discourse Analysis   Any of several methods for studying talk, conversation, and, more broadly, verbal communication. The general approach is to treat utterances not so much as stores of meaning to be deciphered, but more as acts or performances to be interpreted.

Discrete Variable   Commonly, another term for *categorical (or *nominal) variable. Compare *continuous variable.

More formally, a discrete variable is one made up of distinct and separate units or categories. When a variable is discrete, only a finite number of values separates any two points. While all categorical variables are discrete, in some usages there might be dispute about whether to label particular variables discrete or continuous. This matters because it determines appropriate statistical techniques.

For example, the number of people in a family is clearly a discrete variable. So is flips of a coin; if you flip a coin 10 times you can't get 3.27 tails. But the distinction is not always so clear. Take personal income. It looks like a continuous variable, and it is usually treated as one in research. Millions of possible values stretch from zero to Bill Gates's income. More strictly, however, income is discrete. Income does not come in units smaller than one cent; there is only one value between $411.01 and $411.03 ($411.02). Thus, while income is measured on a *ratio scale, it is a discrete variable. By contrast, weight is a truly or a strictly continuous variable. No matter how close two people's weights might be, there is always an intermediate value, although an ordinary scale might not capture it. Because of limits in how accurately we can measure, all measurements are discrete in practice.

Discriminant Analysis (DA)   A form of *regression analysis designed for classification. It allows two or more *continuous *independent variables (or *predictor variables) to be used to place individuals or cases into the categories of a *categorical *dependent variable. DA also provides a means of calculating a weighted combination of all independent variables so as to be able to cut the dependent variable into discrete categories. Also called discriminant "function" analysis. It is called "multiple" discriminant analysis when subjects are to be placed in more than two categories. Compare *logistic regression.

DA was originally used for classification work, such as deciding whether a collection of thigh bones dug up by paleontologists belonged to early hominoids, chimps, or baboons. Continuous variables, such as the bones'

D

length, weight, and circumference, were used to "discriminate" among them and place them in the right categories.

As a second example, illustrating another use of DA, suppose researchers wanted to use data about previous high school students to figure out which current students were and were not likely to graduate. The categorical dependent variable would be graduation yes/no. The continuous predictor variables might be number of days absent, grade point average, score on a verbal ability test, etc. A successful discriminant analysis would enable the researchers to predict, with some accuracy, who would be likely to graduate and who would not, and to compare the relative importance of each of the predictor variables.

Discriminant Function   A combination of the observed *independent variables in a *discriminant analysis that aids in distinguishing (discriminating among) categories of the *dependent variable. Like *factors in a factor analysis and *canonical variates, discriminant functions are ways of dealing with groups of related variables and relating them to other variables. See *structural equation modeling.

Discriminant Validity   A measure of the *validity of a *construct that is high when the construct fails to *correlate with other, theoretically distinct constructs. Discriminant validity is often called "divergent validity" and is the mirror image of *convergent validity.

For example, suppose researchers are writing a questionnaire containing several questions designed to measure the construct "patriotism." They worry that respondents may just be giving the answers they think are "proper" or that they think the researchers want to hear (*social desirability bias). So the researchers include questions that measure the construct "socially desirable responding." If the two measures were not correlated, the measure of patriotism would have more discriminant validity, that is, it would be unrelated to a measure of something to which it should not be related if it were valid.

Disjoint Sets   In *set theory, sets with no common elements, that is, sets that are "joined by" no common elements, such as the set of all males and the set of all females.

Disordinal Interaction   Said of an *interaction effect when the lines on a graph plotting the effect cross. When the lines do not cross, the interaction is called *ordinal. The two graphs that follow show a disordinal and an ordinal interaction. In neither case are the lines parallel. If they were parallel, there would be no interaction. In an ordinal interaction, the lines *would* cross if extended further. The fact that they are not extended to that point could mean that the researcher was not interested in those levels of the variables—in the example, no more than 10 treatments of fertilizer, and of course, by logical necessity, no fewer than zero treatments.

Say a gardener had nine rows of tomatoes with nine plants in each row. She planted three different types of tomato plants, A, B, and C, three rows each. She gave each of the nine "columns" of tomatoes a different number of doses of fertilizer and kept a record of the total weight of tomatoes produced by each type of plant at each dose level. Up to a certain point, fertilizer increases the tomato crop, but, for different types, it does so at different rates, which means there is an interaction effect. The upper figure shows a disordinal interaction; the bottom figure shows an ordinal interaction.

Table D.6   Disordinal Interaction of Plant Type and Fertilizer Dose

Type	Row	0	1	2	3	4	5	6	7	8
		0	*1*	*2*	*3*	*4*	*5*	*6*	*7*	*8*
A	1	a	a	a	a	a	a	a	a	a
	2	a	a	a	a	a	a	a	a	a
	3	a	a	a	a	a	a	a	a	a
B	4	b	b	b	b	b	b	b	b	b
	5	b	b	b	b	b	b	b	b	b
	6	b	b	b	b	b	b	b	b	b
C	7	c	c	c	c	c	c	c	c	c
	8	c	c	c	c	c	c	c	c	c
	9	c	c	c	c	c	c	c	c	c

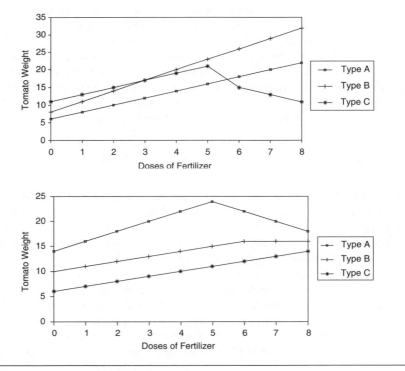

Figure D.4   Disordinal/Ordinal Interaction

Dispersion, Measure of   A statistic showing the amount of *variation or spread in the scores for, or values of, a *variable. When the dispersion is large, the scores or values are widely scattered; when it is small, they are tightly clustered. Two commonly used measures of dispersion are the *variance and the *standard deviation. A measure of dispersion always implies a measure of *central tendency, such as a *mean. For example, the standard deviation measures deviation *from the mean.*

Dissimilarity Index   An alternative to the *Gini index, often used as a measure of neighborhood segregation. The index indicates the percentage of a group that would have to move to achieve a proportional residential pattern in a geographic region.

Distance   Sometimes used in statistics to mean difference, as when the distance between two populations is the difference between their means. See *Mahalanobis distance.

Distribution   A ranking, from lowest to highest, of the values of a *variable and the resulting pattern of measures or scores, often as these are plotted on a graph. Usually either a *probability distribution or a *frequency distribution. See *array, *kurtosis, *normal curve, *sampling distribution, *skewed distribution.

For example, if researchers recorded the closing sale price of all stocks traded on the New York Stock Exchange on a given day and arranged the prices in *ascending order so that they could study patterns in the prices, they would have constructed a distribution.

Distribution-Free Statistics (or Tests)   Another term for *nonparametric statistical tests; so called because they do not require assumptions about the form of the distributions of the *populations from which *samples are drawn. Examples include the *chi-squared and *Wilcoxon tests. See also *bootstrap methods.

Disturbance   Another term for *noise in *information theory; broadly used to mean *random error.

Disturbance Term   See *error term.

Divergent Validity   Another term for *discriminant validity.

Diversity Index   A measure of the degree to which different categories in a *population occur with unequal frequencies. If there is only one category in the population, the value of the index is zero. Compare *Gini index.

Dividend   In division problems, the number that is divided—by the *divisor or *denominator. The part of a fraction that is above the line. Also called the *numerator.

Divisor   A quantity used to divide another quantity (the *dividend). The part of a fraction that is below the line. Also called the *denominator.

DK   Common abbreviation for "Don't know" in survey research.

Domain   (a) A subgroup of a *population that is of special interest to the researchers *sampling it. (b) The content area studied in a *domain-referenced test. (c) In set theory, a set of numbers that can serve as a replacement for a variable; also called "replacement set." See *function.

Domain-Referenced Test   A type of achievement test that measures a person's absolute level of performance in a specific area or "domain," such as long division or economic history. Domain-referenced tests usually measure content areas more specifically defined than other achievement tests. Compare *criterion-referenced test, *norm-referenced test.

Domain Sampling   Sampling items, such as questions on a questionnaire, in a particular subject area or *domain.

For example, a researcher might be interested in the domain of respondents' attitudes toward affirmative action. Rather than study all responses to all questions that are pertinent, the researcher could take a sample of the questions in the domain (which, in this usage, is a "*population" of items).

Dominance Matrix   A table showing winners and losers (or analogous relations) in a *game theory or similar problem.

For example, the following matrix shows the winners and losers in a chess tournament (1 = yes/win, 0 = no/lose). Reading across the rows we can see, for instance, that Anne beat Carl, but lost to Bob and Dina. Or, reading down a column, we can see that Dina lost none of her matches, while Carl lost all of his. See *directed graph for another way to show the same data.

Table D.7   Dominance Matrix of Chess Tournament Results

		Loser			
		Anne	Bill	Carl	Dina
Winner	Anne	—	0	1	0
	Bill	1	—	1	0
	Carl	0	0	—	0
	Dina	1	1	1	—

DOS   Disk Operating System. One of the earliest of the operating systems, that is, computer software that coordinates *software's demands on *hardware. Like other operating systems, DOS makes a kind of "map" of the disks so as to manage, store, and keep track of files. Other operating systems include UNIX, OS, and Windows.

D

Dot Plot A graphic representation of the distribution of one variable only as distinct from the typical scatter plot, which shows the relation of two variables. Each value is represented by one dot. Compare *stem-and-leaf diagram and *bar graph. In the following figure, the number of students earning particular grades is depicted: Six students earned a 1; 20 earned a 2; and so on.

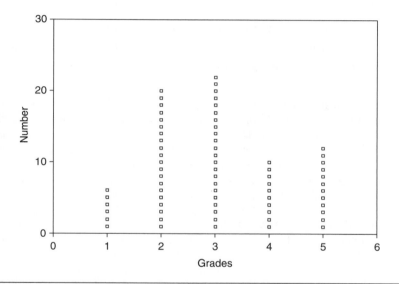

Figure D.5 Dot Plot

Double-Blind Procedure A means of reducing *bias in an *experiment by insuring that both those who administer a *treatment and those who receive it do not know (are "blind" to) which subjects are in the *experimental and *control groups, that is, who is and who is not receiving the treatment.

For example, in a study of the effectiveness of various headache remedies, 80 headache sufferers could be *randomly assigned to four groups. Group A would receive aspirin, Group B ibuprofen, Group C acetaminophen, and Group D a *placebo. The pills might be color-coded, but otherwise look the same so that the experimenter handing them out would not know which subjects were getting which; and, of course, the subjects would not know. When subjects experienced pain, they would be given pills depending upon their group and then asked about the relief they got from their pills. Their responses would be data used to evaluate the

D

effectiveness of the various remedies. If the experiment used true double-blind procedures, the researchers analyzing the data would not know, until after they had reached their conclusions, which group received which remedy; they would only know, say, that on average the blue pills worked better than the red ones.

Double-Tailed Test   Another name for a *two-tailed test and *nondirectional test, that is, one for which the *region of rejection of the null hypothesis is made up of (usually equal) areas at both ends of the *sampling distribution. See illustration at regions of rejection. Compare *one-tailed test/hypothesis.

Download   Transfer data from one computer to another.

_D_ Test   Also called the Kolmogorov-Smirnov D Test. A test of the *statistical significance of the difference between two *frequency distributions. It is a *nonparametric test.

Dummy Coding   A way of *coding *categorical variables such that membership in a category is indicated by a 1 and nonmembership by a 0. So called because the zero is silent ("dumb") about nonmembership. One advantage of dummy coding is that it allows researchers to use statistical techniques that assume *interval-level data on variables measured only at the *nominal or *ordinal levels. Also called "indicator coding." See *dummy variable. Compare *effects coding, *orthogonal coding.

Dummy Table   An empty or blank table (one that "says" nothing) constructed before data are collected and into which the data will be put once they are collected.

Dummy Variable   A *dichotomous variable, usually *coded 1 to indicate the presence of an attribute and 0 to indicate its absence. Example: 1 = female; 0 = not female. This coding facilitates the use of *interval-level statistical techniques, which could be harder to interpret if the variable were coded otherwise, such as, female = 2, male = 1. Also called "indicator variable." See *dummy coding, *multiple classification analysis.

When a variable has more than two categories, a series of dummy variables can be used. For example, say we wanted to use *regression analysis to study the effects of three kinds of growing conditions (A, B, C) on weight (Y) of pumpkins. The coding would be condition A: 1 = yes; 0 = other than A; B: 1 = yes; 0 = other than B; C: 1 = yes; 0 = other than C. The results of weight (Y) by condition for 15 cases put in dummy variable form would appear as in the following table. Upon examining the table, you might well ask: Whatever happened to C? C need not be included because it can be deduced from the coding of A and B. Indeed, it _must_ not be included, since the number of dummies must be one less than the number of categories in order to avoid *multicollinearity.

Table D.8 Dummy Variable Coding

Case	A	B	Y
01	1	0	20
02	1	0	21
03	1	0	23
04	1	0	22
05	1	0	18
06	0	1	15
07	0	1	12
08	0	1	11
09	0	1	19
10	0	1	17
11	0	0	24
12	0	0	26
13	0	0	29
14	0	0	28
15	0	0	29

D

Duncan's Multiple-Range Test   A test used after an *analysis of variance (ANOVA) has been conducted to determine which sample means differ significantly from one another.

Dunn's Multiple Comparison Test   Another term for the *Bonferroni test statistic, that is, a method for multiple comparisons of *treatment effects in *regression analyses and *ANOVA designs. It adjusts (downward) the *alpha level depending on the number of comparisons.

Dunnett's Test   A method of controlling for *Type I errors when *multiple comparisons are made. It compares each of a number of *treatments with a *control. Like other such *post hoc methods (e.g., Tukey's, Duncan's, and Scheffé's), Dunnett's test adjusts the size of the *critical value used to determine whether an observed difference between two means is statistically significant. Compare *Bonferroni technique.

Dunn-Sidak Modification   A revision of the *Bonferroni technique for multiple tests of significance.

Duration Analysis   Another term for *survival analysis. See also *event history analysis.

Duration Recording   Measuring the amount of time a particular behavior lasts, such as using a stopwatch to record how long research subjects spend talking to one another.

Durbin *h* Statistic   A variation on the *Durbin-Watson statistic used to detect *autocorrelation in a *regression in which some of the independent variables are in a *lagged relationship.

Durbin-Watson Statistic   A diagnostic test for *autocorrelation, or serial correlation, in a *time-series, *OLS *regression analysis. The larger the autocorrelation, the less reliable the results of the regression analysis; as the autocorrelation increases, the Durbin-Watson goes down.

DV   Abbreviation for *dependent variable.

Dyad   Two persons interacting; often thought of as the most elementary sociological unit.

Dynamic Model   In economics and related disciplines, a model in which at least one variable is measured over time.

Dysfunction   Any element of a system that hinders the overall operation of the system, as hostility between groups in a society might impede the functioning of the society. See *functionalism.

E (a) Upper- or lowercase *E*, the usual symbol for *error, or *error score. See *residual. (b) Uppercase *E*, *expected value, as in $E(x) = .33$, which means, the expected value of x is .33. (c) Lowercase *e*, the symbol for the "universal constant," or "Euler number" (2.7182818 . . .), which is an *irrational number that is the base of natural *logarithms, and which is used in many calculations, such as figuring compound interest. The formula for this *e* is $1 + 1/1! + 1/2! + 1/3! + 1/4! . . .$ See *factorial; compare *pi. (Note that there is also a Euler constant, gamma.)

Ecological Correlation A correlation between two variables based on grouped data, such as averages for a geographical area or for social groups. One can commit an *ecological fallacy by using such correlations to draw conclusions about individuals.

   For example, the correlation between the gross national income for various nations and average education level for those nations would be an ecological correlation. Such a correlation would not, however, be valid evidence for an individual to use in deciding whether she should go back to school in hopes of earning a higher income.

Ecological Fallacy An error of reasoning committed by coming to conclusions about individuals based only on data about groups. Compare *Simpson's paradox, in which individual-level associations are changed by aggregation.

   For example, if crime rates were higher in areas with a high concentration of elderly people, you would be committing an ecological fallacy if you concluded that elderly individuals are more likely to commit crimes.

   Reasoning in the opposite direction, from data about a few individuals to generalizations about groups, is also a widespread form of fallacious thinking, sometimes called *fallacy of composition.

E

Ecological Inference Problem   The difficulties in avoiding *ecological fallacies are sometimes referred to in this way, usually by those who do not think that reasoning from data about groups to conclusions about individuals is necessarily fallacious in all cases.

Ecological Validity   (a) A kind of *external validity referring to the generalizability of findings from one group to another group. Usually used when a study does not meet the criteria. Compare *population validity. (b) The extent to which a measurement taken in an experiment or on a survey reflects what subjects do in real life.

For example (definition a), studies of 19-year-old college students might not be generalizable to 19-year-olds who are not attending college. If a study made such generalizations from one group or context to the other, it could be lacking in ecological validity.

An example of definition (b) would be having subjects rate each of the four candidates in the upcoming primary election on a scale of 1 to 10. Having them pick one candidate would be more ecologically valid, because that is what they will have to do on election day.

Econometrics   (a) The application of statistical methods to economic data, usually to forecast economic trends and decide among policies. (b) The branch of economics applying statistical *models, often models based on multiple *regression, to economic problems.

EDA   Abbreviation for *exploratory data analysis.

Effect   (a) Broadly, a phenomenon believed to have been caused, influenced, or determined by another phenomenon, as when inflation is held to be the effect of declining productivity. (b) In *analysis of variance (ANOVA), effect refers to differences among group *means, differences presumably caused by treatments received by the groups. *Main effects are differences among group means for levels of a *variable (*factor) apart from the effects of other variables. *Interaction effects occur when the effect for one factor (variable) differs depending on the levels of another factor. See the example at *disordinal interaction.

Effects are often discussed as *total* effects, often measured by *zero-order correlation. Total effects are composed of *direct* effects, that is, those which are not mediated or which remain after the effects of *mediating variables have been removed; and *indirect* effects, that is, effects mediated or transmitted by *intervening variables. Direct effects plus indirect effects add up to total effects.

Effect Coefficient   In *path analysis, the total effect (that is, direct plus indirect effects) of an *independent variable on the *dependent variable.

Effectiveness   Ability to achieve goals well; or the degree to which intentions are achieved. Often contrasted in *evaluation research with *efficiency, which is a measure of cost relative to output.

E

Effective Procedure   A series of steps that work in solving problems; an
*algorithm.

Effect Modifier   A characteristic of research subjects that interacts with a
*treatment; thus another term, and a clearer one, for *moderating variable.
For example, if a treatment had stronger (or weaker) effects for young than
for old people, age would be an effect modifier. The term is more often used
in epidemiology than in the social sciences.

Effects Analysis   Also known as *decomposition of effects. Trying to determine what the relation between two variables would be with prior or
*exogenous variables controlled. See *partial correlation, *elaboration, and
the final paragraph of *effect.

Effects Coding   Also called effect coding. A way of coding *categorical
variables in a *regression analysis. It uses 1, 0, and −1, unlike *dummy
coding, which uses only 1 and 0. Effect coding gets its name from the fact
that when it is used, the *regression coefficients (betas) show the effects of
the treatments. Compare *orthogonal coding.

Effect Size (ES)   (a) Broadly, any of several measures of association or of the
strength of a relation, such as *Pearson's r or *eta. ES is often thought of as a
measure of *practical significance. (b) A statistic, often abbreviated d, Cohen's
d, *D, or delta, indicating the difference in outcome for the average subject
who received a *treatment from the average subject who did not (or who
received a different level of the treatment). This statistic is often used in
*meta-analysis. It is calculated by taking the difference between the control
and *experimental groups' means and dividing that difference by the *standard deviation of the control group's scores—or by the standard deviation of
the scores of both groups combined. In psychological research it is often
referred to as *the* effect size statistic, but it is in reality one of many. (c) In statistical *power analysis, ES is the degree to which the *null hypothesis is false.

Efficiency   (a) In *research design, said of a procedure that uses fewer
resources for the same results or that gets more results using the same
resources. (b) In statistics, a property of an *estimate of a *population
*parameter; the better the estimate, the greater the efficiency. Efficiency is
a measure of the *variance of an estimate's *sampling distribution; the
smaller the variance, the better the estimator. Compare *effectiveness.
(c) In economics, a measure of cost per unit of output; the lower the cost
per unit, the higher the efficiency.

Efficient Estimator   See *efficiency, definition (b), and *estimator. Among
*unbiased estimators, the one with the smallest *variance is called the
"best" or "most efficient" estimator. See *BLUE.

Eigenvalue   A statistic used in *factor analysis to indicate how much of the
variation in the original group of variables is accounted for by a particular

factor. It is the sum of the squared *factor loadings of a factor. Eigenvalues of less than 1.0 are usually not considered significant. Usually symbolized lambda [Λ]. Also called "characteristic root" and "latent root."

Eigenvalues have similar uses in *canonical correlation analysis and *principal components analysis.

Elaboration   A process of studying *correlations between *variables by observing how they are affected when *controlling for the effects of other, *intervening variables. One goal of elaboration is to uncover *spurious correlations. Compare *effects analysis.

Elasticity   Percentage change in one *variable divided by percentage change in another variable. Used most often in economics.

Element   (a) In *set theory, any one of a set's members. (b) In *matrix algebra, an individual value. See *vector.

Elementary Event   See *event, elementary.

Elementary Unit   Another term for *unit of analysis, the point at which one analyzes (breaks down) the subject matter no further. For example, in a study of cities or of families and their characteristics, cities and families would be the elementary units, not the individuals living in them.

Emic   *Culturally relative approaches to the study of anthropology that stress participants' understanding of their own culture. Derived, by an indirect route, from the linguistics term phon*emic*. Usually contrasted with *etic.

Empirical   Said of *data based on observation or experience and of findings that can be verified by observation or experience. Often contrasted with "theoretical." Compare *deductive, *objective.

Empirical Bayes Procedure   Estimation or prediction based on *Bayesian inference when the *prior distribution is based on data rather than subjective judgment.

Empirical Generalization   A statement about observable regularities made without an attempt at explanation. Such factual statements can sometimes be useful, but without being explained by a *theory, they usually add little to science. Compare *middle-range theory.

An example of an empirical generalization is: "men in their twenties have an unusually high rate of killing themselves and others in automobile accidents." There are many possible explanations. Say we theorized that young men are socialized to a subculture that defines traditional manhood as reckless disregard for personal safety. If this theory is true, it could be used to explain other aspects of young men's behavior (such as smoking, drinking, or participation in violent sports). Furthermore, with our theory about traditional male culture, we could make predictions about other groups. For example, as women became more integrated into male-dominated society, their auto death rate should go up (along with their drinking and smoking rates). Or we might predict that the auto death rate for men somewhat outside traditional male culture (such as homosexual men) would be lower.

Empiricism   Any approach to research relying heavily on observation and experiment; also, the belief that only such an approach yields true knowledge. Compare *positivism, *idealism.

Empty Cells   A problem in research using *cross tabulations that arises from having too many categories or too few subjects. Whenever a category has no subjects that fit into it, you have an empty cell. See *cell.

For example, suppose you survey a *sample of 100 people on their attitudes. You are interested not only in overall response of the sample, but also in the attitudes of different groups of people. Among the groups you think are important are gender, age, race, education, and occupation. It is clear that with only 100 people answering your survey you would not have many people in each category. Some will almost surely be empty, for example, white female blue-collar workers over 60 with more than 12 years of education. Of course, you might have several people in that category, but if you do, you will be short of people in other categories.

Empty Set   A *set that contains no *elements. Also called "null set."

Encode   To put into a *code, as by assigning numbers to categories. Encoding always involves simplifying observations. See *coding.

Endogenous Variable   A *variable that is caused by other variables in a causal system. Generally contrasted with *exogenous variable. See *path diagram. In the following figure, Child's Aspirations, Child's Education, and Child's Income are endogenous. Parents' Education and Parents' Aspirations are exogenous.

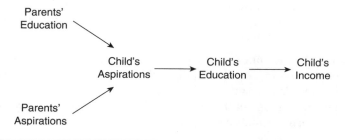

Figure E.1   Endogenous Variable

Entails   Said of a statement that must logically or inevitably follow from another statement. "A entails B" means that if A is true, B must also be true. Such entailment is very rare in the social and behavioral sciences. Compare *necessary and *sufficient conditions.

Epidemiology   The study of health problems in a population, especially infectious disease. Many research issues in epidemiology are parallel to research issues in the social sciences, especially problems that arise from using nonexperimental designs to make causal inferences.

Epistemology   Literally, "the study of knowledge." That branch of philosophy concerned with the nature and criteria of knowledge. Methodological debates in the social and behavioral sciences are often the result of differences of opinion about epistemological issues and often raise epistemological questions. See *idealism, *empiricism, *materialism, *rationalism, *realism.

EPSEM   *Equal probability selection method.

Epsilon [E, ε]   Symbol for the *random error component in a *regression equation.

Epsilon Squared   Another term for *adjusted R^2.

EQS   A statistical program designed for *structural equation modeling. Compare *LISREL.

Equality of Variances   Also called "homogeneity of variance" and *homoscedasticity. A basic assumption of several statistical tests (e.g., *t-test, *ANOVA) which, when violated, may require transforming the data or using a *nonparametric test. See *heteroscedasticity for a graphic illustration.

Equal Probability Selection Method (EPSEM)   Any *sampling procedure in which all members of the *population have an equal probability of being included. Compare *probability sample.

Equation   A formal statement that two mathematical expressions, placed on either side of an equal sign (=), are equal, such as: $12 \times 9 = 108$, or $Y' = bX + a$. Compare *model, *function.

Equivalence, Coefficient of   A measure of *reliability used for two or more forms or versions of a test to determine the extent to which they yield the same (or equivalent) results. It is a *correlation between the two forms and is a widely used gauge of the reliability of *standardized tests. See *reliability coefficient, *Cronbach's alpha.

ERIC (Educational Resources Information Center)   Managed by the National Institute of Education, ERIC indexes and abstracts journal articles (in Current Index to Journals in Education, or CIJE) and other documents and research reports (in Resources in Education, or RIE).

Error (a) The difference between a true value and an observation, measurement, or estimate of that value. (b) The difference between a predicted or estimated score and an observation; this is often called a *deviation score. Symbolized as e or E or *epsilon. Note that in statistics, error usually does not mean mistake; rather it means imprecision. But see *Type I error, which is a mistake.

Error of the First Kind Another term for *Type I error.

Error of the Second Kind Another term for *Type II error.

Error Score The difference between an estimated value, such as a *mean, and the actual value of a subject on a variable. Also called *deviation score.
For example, say that the average (mean) verbal GRE score for students in your department were 560 and your score was 580. If the head of your department used the mean score to predict your score, the error score for you would be −20 (560 − 580 = −20).

Error Sum of Squares In *analysis of variance, the *within-group sum of squares, that is, the part not explainable by the *treatment effects. Also called "residual sum of squares." See *error variance.

Error Term The part of an *equation indicating what is unexplained by the *independent variables. The error term specifies how big the unexplained part is. Also called the *residual term, since it is what is left over after one subtracts from the total *variance in the *dependent variable the part that can be explained by the independent variables. When subjects in an experiment are randomly assigned, differences among them become part of the error term. Also called "disturbance term."

Error Variance Any uncontrolled or unexplained variability, such as *within-group differences in an *ANOVA. Also called *random error, "random variance," and *residual. The error variance is the variance of the *error term.

ES *Effect size.

Essentialism Used pejoratively to indicate an erroneous belief that one can accurately describe the essence of persons or things. Often used as a synonym for racism or sexism. The term has its roots in older discussions in *epistemology concerning the status of *concepts and their relationship to things. Some see *falsificationism as a position opposed to essentialism.

Estimate The value of an *estimator.
For example, we might use the *mean daily caloric intake of a *sample of adults to estimate the mean daily caloric intake of a *population of

E

adults. The sample mean is our estimator; if we calculated it to be 2,570, then 2,570 would be the estimate given us by our estimator.

Estimation   Using a *sample *statistic to determine the probable value of a *population *parameter. See *inferential statistics. Estimation is used when the researcher is not testing a hypothesis. The size of a relation, rather than its *statistical significance, is the goal in estimation. In other words, in estimation the researcher has no hypothesis about the population parameter, but in hypothesis testing the researcher must have one.

For example, let's say you wanted to know the average (*mean) verbal SAT score of students at your university. Rather than going through the files and getting the *data for several thousand undergraduates, you could take a *random sample of, say, 200 files. Suppose the mean verbal SAT score of those 200 students was 553. If you used that average score to estimate the score of the population of all students, this would be a *point estimation. If you said that the interval between 513 and 593 was likely to contain the mean score of the population, this would be *interval estimation.

Estimator   (a) Any formula that can be used to make an *estimate. (b) A *sample statistic that is used to determine a probable value of a *population parameter, as one might use a known *mean value of a sample to estimate the value of the population mean. See *estimate, *estimation. Good estimators should be consistent, *unbiased, and *efficient.

Eta [H, η]   A *correlation coefficient that does not assume that the relation between two variables is *linear. Thus, it can be used to express a *curvilinear relationship. It can be used only with categorical independent variables and a continuous dependent variable (as in ANOVA). Eta is read in the same way as other correlation coefficients. Also called *correlation ratio. Compare Pearson's r. See *polynomial regression analysis.

Eta Squared   A measure of how much of the *variance in a *dependent variable (measured at the *interval level) can be explained by a *categorical (nominal, discrete) *independent variable. It may also be used as a measure of association between two interval variables. Eta squared can be interpreted as a *PRE measure, that is, it tells us how much better we can guess the value of the dependent variable by knowing the independent variable. Eta squared in *ANOVA is analogous to *R^2 in multiple regression; it is an estimate of the variance associated with all the independent variables taken together.

Ethics   Principles of good conduct. In the context of research, ethics usually refers either to the protection of subjects or to honesty in reporting results.

Ethnographic Research   Any of several methods of describing social or cultural life based on direct, systematic observation, such as becoming a participant in a social system. Ethnographic methods are most commonly used by anthropologists. See *descriptive research, *participant observation.

E

Ethnomethodology   A type of *ethnographic research in sociology founded by Harold Garfinkel. It focuses on the commonsense understanding of social life held by ordinary people (the "ethnos"), usually as discovered by *participant observation. Often the observer's own methods of making sense of the situation become the object of investigation.

Ethology   Research methods stressing observation and detailed descriptions of behavior in natural settings. The term originally referred to the study of animal behavior, but it has come to be used in research on human behavior when the methods are strongly observational and minimally interpretive. See *ethnographic research.

Etic   Methods of study in anthropology stressing material—rather than cultural—explanations for social and cultural phenomena. Derived, by an indirect route, from the linguistics term phon*etic*. Compare *emic, with which etic is usually contrasted.

For example, an anthropologist using an etic approach might look for the origins of cannibalism in a need for protein; her emic-minded colleague might stress the religious needs that cannibalistic rituals satisfied.

Etiology   The study of causes; usually but not exclusively causes of diseases.

Euler's Constant   An *irrational number, usually symbolized gamma, with a value of .577216 . . .

Evaluation Research   Investigations, using any of several methods, designed to test the *effectiveness or impact of a social program or intervention. It is very often conducted by interdisciplinary teams of researchers. Evaluation research is a type of *applied research. Also called "program evaluation." Compare *assessment research.

Evaluation research became important in the 1960s with the expansion of social welfare programs in that decade and has become a routine requirement of federal programs in the United States.

Examples of evaluation research include studies designed to tell whether a school desegregation plan improved intergroup harmony, whether new sentencing guidelines deterred crime, or whether driver education reduced fatal accidents.

Event   In *set theory, any group of *elementary events. A *subset. Also called "event class" or "compound event" to distinguish it from an elementary event.

For example, if the elementary event were the 7 of clubs in a deck of cards, an event would be the 7s, the clubs, or the black cards.

Event, Compound   An *event made up of two or more events. Either a *union or *intersection of sets is a compound event.

For example, hiring an employee who is highly motivated *and* talented is a compound event.

E

Event, Elementary   In *set and *probability theory, a single *data point in, or element or member of, a *sample space. Also called "simple event."

For example, the 7 of clubs would be an elementary event in a deck of cards (the sample space). A particular college student would be an elementary event in the *population (or sample space) of all college students.

Event History Analysis   Methods for studying the movement over time of subjects through successive states or conditions by asking them to remember biographical data (an alternative approach uses *archives containing longitudinal data). The goal of the research is to study change from one state to the next and how long each of the states lasts. "Events" are changes from one categorical state to another. For example, one could use event history analysis to study marital status, with the states or conditions being: unmarried, married, divorced, remarried, widowed. See *longitudinal study, *time-series data, *survival analysis.

By contrast, *panel studies, which also study subjects over time, use successive *waves of surveys or interviews. Since they do not usually investigate what happened between the waves, panel studies can be thought of as a series of *cross sections of the same group.

Event Space   Another term for *sample space.

Evidence-Based Practice   Said of practice in medicine, and by extension in other fields such as education, that is based on the systematic review of the most current research in the field.

Exact Test   See *Fisher's exact test.

Exhaustive   Said of a group of conditions, events, or values of a variable that, when taken together, account for (or "exhaust") all possibilities.

For example, age categories 0–19, 20–39, and 40-plus are exhaustive; everybody fits into one of them. On the other hand, Christian, Hindu, Islamic, and Jewish is not an exhaustive list of religious affiliations. It could be made so, however, by adding Other and None to the list. Compare *mutually exclusive.

Exogenous Variable   A variable entering from and determined from outside the system being studied. A causal system says nothing about its exogenous variables. Their values are given, not analyzed. Also called "prior variables." See *endogenous variable. In *path analysis, cause is illustrated by an arrow →. If a variable does not have an arrow pointing at it, it is exogenous.

For example, say we were studying the relation of hours spent practicing to score in an archery contest. Subjects' dexterity and strength might be related to their scores, but would be exogenous variables for the purposes of our study.

Expected Frequency In *contingency table problems, the frequency you would predict ("expect")—if you knew only the *marginal totals and if you assumed the *independence of the variables. With the *chi-squared test, the expected frequency is the predicted value when the *null hypothesis is true.

For example, suppose that in a *sample of 100 adults you had 60 women and 40 men, and that 70 of the 100 adults had graduated from high school and 30 had not. You could put the results in a contingency table as in the following:

Table E.1 Expected Frequency

	Men	*Women*	*Total*
Graduates	a	b	70
Nongraduates	c	d	30
Total	40	60	100

Given this information, you can compute the expected frequency for each of the cells a, b, c, and d. To find the expected frequency for a cell, multiply its row marginal (total) by its column marginal (total) and divide the result by the total number of subjects. For example, the expected number of women high school graduates (cell b) would be $60 \times 70 = 4200/100 = 42$. You would expect 42 of your sample to be female high school graduates. If your expected frequencies were significantly different from the actual, observed frequencies (you could determine this with a *chi-squared test), you could conclude that there was probably some relationship between the variables such that one sex was more likely to have graduated from high school.

Expected Value (a) The *mean value of a variable in repeated samplings or trials. (b) The mean of the *sampling distribution of a statistic. Usually symbolized E, as in $E(X)$, which means the expected value of X.

The idea grew out of gamblers' calculations of how much they could expect to win (or lose) in a fair game, in the long run, with a bet of a certain size. Say you play roulette, making 1,000 bets of one dollar on your favorite number. Each time you win you get $35; each time you lose, the croupier takes your dollar. Your odds of winning on most wheels are 37 to 1, which means that in the long run you will lose about $55 for every 1,000 bets. If you tried the experiment once, you might do considerably better or worse. But if you tried it many times, your average result for each 1,000 bets would be to lose about $55. The more times you made the 1,000 bets, the closer your average would get to the expected value of a $55 loss.

Note that the expected value is not necessarily the most common (modal) value. It can even be an impossible value. If, for example, a

E

variable can have a value of either 1 or 2, the expected value is 1.5, a value which never occurs.

Experience Sampling Method   Any of several methods that require subjects to record what they are doing or thinking or how they are feeling at times specified by a researcher. For example, subjects might be asked to enter data into a log when beepers they were wearing sounded.

Experiment   (a) In behavioral research, a study undertaken in which the researcher has control over some of the conditions of the study and control over (some aspects of) the *independent variables being studied. The independent variables are manipulated rather than only observed. *Random assignment of subjects to *control and *experimental groups is usually thought of as a necessary criterion of a true experiment. Compare *natural experiment, *quasi-experiment, *secondary analysis, *descriptive research.

(b) In probability theory, an experiment is an act or a process that leads to a single outcome (an elementary *event) when that outcome cannot be predicted with certainty, such as flipping a coin or asking potential customers whether they would consider buying a particular product.

For example, if you interviewed moviegoers as they exited a theater to see if what they saw influenced their attitudes, this would not be a *true experiment (definition a); you had no control over who the subjects were or what film they watched or the conditions under which they watched it. On the other hand, if you chose a room, a film, and subjects to assign randomly to control and experimental groups and interviewed these subjects about the effects of the film on their attitudes, that approach would be more experimental.

Experimental Design   The art of planning and executing an *experiment (definition a). The greatest strength of an experimental research design, due largely to *random assignment, is its *internal validity: One can be more certain than with any other design about attributing *cause to the *independent variables. The greatest weakness of experimental designs may be *external validity: It may be inappropriate to generalize results beyond the laboratory or other experimental context. See *research design.

Experimental Error   Differences in results among experiments repeated using identical procedures. When experiments are repeated, the results are rarely if ever exactly the same—even if the experiment is well designed and the experimenters make no mistakes. Compare *error term, *random variation.

Experimental Group   A group receiving some *treatment in an experiment. Data collected about subjects in the experimental group are compared with data about subjects in a *control group (who received no treatment) and/or another experimental group (who received a different treatment).

E

Experimental Unit The smallest independently treated unit of study. Compare *unit of analysis.

For example, if 90 subjects were randomly assigned to three *treatment groups, the study would have three experimental units, not 90.

Experimenter Effect (a) A type of *confounding effect that occurs when different experimenters working on the same experiment administer different *treatments or *conditions. (b) Any bias introduced by experimenters' expectations. Compare *Pygmalion effect.

For example, say Al, Betty, and Chuck were running an experiment on subjects' reaction time as influenced by three types of visual cues (A, B, and C). If Al always administered cue A, Betty cue B, and Chuck cue C, it would be impossible to tell if differences in subjects' reaction times were due solely to differences in the cues or in part to the way the experimenters administered them. Al, Betty, and Chuck should rotate. Compare *counterbalancing.

Experimentwise Error Another term for *familywise error, that is, increasing the probability of *Type I error by making *multiple comparisons.

Expert System A computer *program modeled on the experience of human experts in decision making and problem solving. These forms of *artificial intelligence have been used in such diverse fields as medical diagnosis and investment banking.

Explained Variance Variance in the *dependent variable that can be accounted for by (statistically associated with) variance in the *independent variable(s). Contrast *error variance.

Explanatory Research Research that seeks to understand variables by discovering and measuring causal relations among them. Generally used to describe *experimental versus *correlational, *exploratory, or *descriptive research designs. Often contrasted, especially in discussions of *regression analysis, with *predictive research.

Explanatory Variable An *independent variable, especially in *regression analysis. Also called *predictor variable and "regressor."

Exploratory Data Analysis Any of several methods, pioneered by John Tukey, of discovering unanticipated patterns and relationships, often by presenting quantitative data visually. The *stem-and-leaf display and the *box-and-whisker diagram are well-known examples. Compare *hypothesis testing, *data mining.

Exploratory Factor Analysis *Factor analysis conducted to discover what *latent variables (factors) are behind a set of variables or measures. Generally contrasted with *confirmatory factor analysis, which tests theories and hypotheses about the factors one expects to find.

Exploratory Research   Said of research that looks for patterns, ideas, or hypotheses, rather than research that tries to test or confirm hypotheses.

Exponent   A superscript symbol written above and to the right of another symbol to indicate how many times it should be multiplied by itself. For example, 7^3 means $7 \times 7 \times 7 = 343$. See *power.

Exponential Growth   Loosely, very rapid growth; as during an economic crisis, the rate of inflation might double every month from 2% to 4% to 8% to 16% to 32% to 64% from January to July.

Exponential Smoothing   Statistical techniques in *time-series data used to give more *weight to more recent data. Exponentially weighted *moving averages are used to accomplish this.

Ex Post Facto Explanation (or Hypothesis)   An explanation about what the facts "will" look like, but which is offered after they have been collected. This is legitimate in *exploratory research. In other circumstances it can be a dubious practice.

Ex Post Facto Research Design   (a) Any investigation using existing data rather than new data gathered specifically for the study. This means that causes will be studied after (post) they have had their effect. (b) Any *non-experimental research design that takes place after the conditions to be studied have occurred, such as research in which there is a *posttest, but no *pretest. Researchers often try to compensate for the lack of a pretest or *baseline data by *matching subjects or otherwise *controlling for variables that might have influenced outcomes. See *case-control study.

Externalities   Unintended, incidental, or external outcomes or effects—usually economic effects. For example, one of the externalities of a honey farm is that the bees will pollinate neighboring plants. An externality of a new shopping mall in a town could be increased traffic congestion.

External Reliability   See *reliability.

External Validity   The extent to which the findings of a study are relevant to subjects and settings beyond those in the study. Another term for *generalizability.

Extraneous Variable   Any condition not part of a study (that is, one in which researchers have no interest) but which could have an effect on the study's *dependent variable. (Note that in this context, extraneous does not mean unimportant.) Researchers usually try to *control for extraneous variables by experimental isolation, by randomization, or by statistical techniques such as *analysis of covariance. Sometimes called "nuisance variable."

Extraneous Variance   Variance caused by an *extraneous variable.

Extrapolation   Inferring values by projecting *trends beyond known evidence, often by extending a *regression line. Compare *interpolation.

Suppose, for example, that you had some measures of the daily high temperatures for a week in June, as listed in Table E.2. If you used these data to guess that the temperature on Sunday would be 84, that would be an extrapolation. If you guessed that Wednesday's temperature had been 76, this would be an interpolation, which is generally a safer kind of inference.

Table E.2   Extrapolation of Temperature

Mon.	Tues.	Wed.	Thurs.	Fri.	Sat.	Sun.
72	74		78	80	82	

Extreme Outlier   See *outlier, *box-and-whisker diagram.

Extrema   *Extreme values.

Extreme Values   The largest and smallest values in a distribution of values. See *range. Compare *outlier.

F   (a) Uppercase *F*, the statistic that is computed when conducting an *analysis of variance. See *F distribution, *F ratio. (b) Lowercase italicized *f*, the usual symbol for *frequency in a table depicting a *frequency distribution. (c) Lowercase *f*, the symbol for *function, as in $Y = f(X)$. (d) Lowercase *f*, symbol for *sampling fraction.

Face Validity   Logical or conceptual validity; so called because it is a form of validity determined by whether, "on the face of it," a measure appears to make sense. In determining face validity, one often asks expert judges whether the measure seems to them to be valid. See *prima facie evidence.

Fact   A piece of information believed to be true or to describe something real. Compare *objective.

Factor   (a) In *analysis of variance, an *independent variable. (b) In *factor analysis, a cluster of related variables that are a distinguishable component of a larger group of variables. See also *latent variable. (c) A number by which another number is multiplied, as in the statement: Real estate values increased by a factor of three, meaning that they tripled. (d) In mathematics, a number that divides exactly into another number. For example, 1, 2, and 4 are factors of 8, since when you divide each of them into 8, the result (quotient) is a whole number. See *factoring.

Factor Analysis (FA)   Any of several methods of analysis that enable researchers to reduce a large number of *variables to a smaller number of variables, or *factors, or *latent variables. A factor is a set of variables, such as items on a survey, that can be conceptually and statistically related or grouped together. Factor analysis is done by finding patterns among the variations in the values of several variables; a cluster of highly intercorrelated variables is a factor. *Exploratory factor analysis was the original

117

type. *Confirmatory factor analysis developed later and is generally considered more theoretically advanced. *Principal components analysis is sometimes regarded as a form of factor analysis, although the mathematical models on which they are based are different. While each method has strong advocates, the two techniques tend to produce similar results, especially when the number of variables is large.

For example, factor analysis is often used in survey research to see if a long series of questions can be grouped into shorter sets of questions, each of which describes an aspect or factor of the phenomena being studied. See *index, definition (c).

Factor Equations   In *factor analysis, equations analogous to *regression equations describing the regression of observed (*manifest) variables on unobserved (*latent) variables. Factor equations have no *intercept, or rather, the intercept is fixed at zero.

Factorial   Said of a whole number (positive integer) multiplied by each of the whole numbers smaller than itself. It is usually indicated by an exclamation point. Factorials are used extensively when calculating probabilities because they indicate the number of different ways individuals or other units of analysis can be ordered. See *permutation.

For example, 5 factorial, or 5!, means: $5 \times 4 \times 3 \times 2 \times 1 = 120$. Therefore, 5 individuals can be ordered in 120 different ways.

Factorial Experiments or Designs   Research designs with two or more *categorical *independent variables (*factors), each studied at two or more *levels. The goal of factorial designs is to determine whether the factors combine to produce *interaction effects; if the treatments do not influence one another, their combined effects can be gotten simply by studying them one at a time and adding the separate effects. See *analysis of variance, *main effect.

For example, a study of the effects of a drug administered at three doses or levels (factor 1) on male and female (factor 2) subjects' psychological states would be a factorial design. There would be an interaction effect between the drug and sex if, say, at some level(s) the drug had a different effect on men and women.

Factorial Plot   A graph of two or more *independent variables (factors) in which nonparallel lines for the different factors indicate the presence of *interaction. See *disordinal interaction and *interaction effect for examples.

Factorial Table   A table showing the influence of two or more *independent variables on a *dependent variable. See *crossbreak for an example.

Factoring   Breaking a number into its *factors, that is, breaking it into parts which, when multiplied together, equal the number.

F

For example, 24 can be factored into 2×12, 3×8, and 4×6. Each of these numbers (2, 3, 4, 6, 8, and 12) is a factor of 24.

Factor Loadings   The *correlations between each *variable and each factor in a *factor analysis. They are analogous to *regression (*slope) coefficients. The higher the loading, the closer the association of the item with the group of items that make up the factor. Loadings of less than .3 or .4 are generally not considered meaningful. See *factor equation, *structural coefficient.

Factor Pattern Matrix   In *factor analysis, a table showing the *loadings of each variable on each factor. The *pattern* matrix is used when the *rotation is *oblique. See *factor structure matrix.

Factor Rotation   Any of several methods in *factor analysis by which the researcher attempts (by the *transformation of *loadings) to relate the calculated factors to theoretical entities. This is done differently depending upon whether the factors are believed to be correlated (oblique) or uncorrelated (orthogonal).

Factor Score   The score of an individual on a factor after the factor has been identified by *factor analysis.

Factor Structure Matrix   In *factor analysis, a table showing the *loadings of each variable on each factor. The *structure* matrix is used when the *rotation is *orthogonal. See *factor pattern matrix.

Failure   In *probability theory, one of the two ways a *Bernoulli trial can turn out.

For example, in flipping a coin, tails might be called *success and heads "failure." While the designations are mostly arbitrary, the term failure is reserved for occasions when the predicted event does not occur.

Fail Safe N   In *meta-analysis, the number of studies confirming the *null hypothesis that would be necessary to change the results of a meta-analytic study that found a statistically significant relationship.

Fallacy of Composition   An error of reasoning made by assuming that some trait or characteristic of an individual must also describe the individual's group. Compare *ecological fallacy.

For example, if you knew one or several white males who were unsympathetic to affirmative action, it would be a fallacy of composition to assume that all white males were equally unsympathetic.

Fallibilism   The philosophical doctrine, developed by C. S. Peirce, to the effect that all knowledge claims, scientific or otherwise, are always fallible or open to question. See *falsificationism, *Heisenberg's uncertainty principle.

False Alarm   See *signal detection theory.

False Negative   Said of a test that wrongly indicates the absence of a condition—for example, the test shows that you don't have cancer when in fact you do. Originating in medical testing, the term has broader use today. Compare *false positive, *Type II error, *specificity.

False Positive   Said of a test that wrongly indicates the presence of a condition—for example, the test shows that you do have cancer when in fact you do not. Originating in medical testing, the term has broader use today. Compare *false negative, *Type I error, *sensitivity.

Falsifiability   Said of theories that can be subject to tests that could prove them to be false. Those who believe in *falsificationism contend that only falsifiable theories are truly scientific.

Falsificationism   The doctrine, originating with Karl Popper, that we can only refute ("falsify") theories; we can never confirm them. A good theory is one that we have tried—repeatedly, but unsuccessfully—to disprove or falsify. Compare *null hypothesis.

Familywise Error Rate   The probability that a *Type I error has been committed in research involving *multiple comparisons. "Family" in this context means group or set of "related" statistical tests. Also called "experimentwise error."

   For example, if you set your *alpha level at .05 and make three comparisons using the same data, the probability of familywise error is roughly .15 (.05 + .05 + .05 = .15). One way around this problem is to lower the alpha level to, say, .01. But this increases the probability of *Type II error. A better alternative is the *Scheffé test. See *Bonferroni technique, *post hoc comparison.

Fan Spread   When the scores or values for two groups grow further apart with the passing of time, plotting this on a graph results in a figure resembling a fan and is called a fan spread. See *interaction effect.

FASEM   Factor Analytic Structural Equation Modeling. See *structural equation modeling, *analysis of covariance structures, *LISREL. The proliferation of names for identical or highly similar techniques is due to the relative newness of the techniques and, perhaps, to the desire of *software companies to come up with distinctive brand names.

F **Distribution**   A *theoretical distribution used to study *population *variances. It is the distribution of the *ratio of two *independent variables, each of which has been divided by its *degrees of freedom. See *F ratio, *chi-squared distribution. The distribution is perhaps most widely used in or closely associated with *analysis of variance, where it is used to assign *p values. Named after Sir R. A. Fisher.

F

Field (a) A place for doing research that is not a laboratory or library, but a naturally occurring set of phenomena, such as insect life in a meadow or a human interaction in a social organization. (b) In a computer program, a place for a specific piece of information in a *data record, usually marked by a column. *Variables are arranged by column or field, cases by row or record. See *data record for an example.

Field Experiment An *experiment in a natural setting (the "field"), not in a laboratory. The researcher in a field experiment does not create the experimental situation, but he or she does manipulate it. School classrooms are a common location for field experiments. See *quasi-experiment, *natural experiment.

Field Notes A record, usually written, of events observed by an *ethnographic researcher. The notes are taken as the study proceeds. Because field notes will later be used for analysis, they should be as close to comprehensive (perhaps even stenographic) as possible.

Field Research (or Study) Research conducted in a real-life setting, not in a laboratory. The researcher neither creates nor manipulates the subject of study, but rather observes it. See *field experiment, *participant observation, *ethnographic research.

FIFO See *first in-first out.

File Any *program or *data set in a computer's memory.

File Drawer Problem In literature reviews and *meta-analyses, a problem of *validity that arises because study results that are not *statistically significant may often go unpublished; rather they are put in researchers' file drawers and are unavailable for review and analysis. This would *bias any meta-analysis, which would tend to overestimate the proportion of research containing statistically significant findings. See *publication bias.

FILO See *first in-last out.

Filter In the context of *time-series analysis, a procedure for converting one time series into another. The most widely known example is the *moving average.

First In-First Out (FIFO) A rule determining the order in which data are processed by a computer; often used in controlling business inventories. Also called "queue processing."

First In-Last Out (FILO) A rule determining the order in which data are processed by a computer; often used in controlling business inventories. Also called "stack processing."

First-Order Interaction Effects Said of the *interaction of two *independent variables with a dependent variable. Second-order interaction effects

F

take place among three independent variables. Higher orders are possible, but are difficult to interpret.

First-Order Partial   Said of a *partial correlation or a *partial regression coefficient that *controls for the (*linear) effect of one *independent variable. A second-order partial controls for two, a third-order for three, and so on. A *zero-order correlation controls for no other variables.

Fisher's Exact Test   A *test statistic for measures of *association that relate two *nominal variables. It is used mainly in 2×2 frequency tables, when the *expected frequency is too small to trust the use of the *chi-squared test. Despite its name, it is no more precise or "exact" than other tests. See *phi coefficient, *Yates's correction.

Fisher's LSD Test   A test of statistical significance used in *post hoc or *multiple comparisons. LSD is short for least significant difference. See *Scheffé test.

Fisher's Z   (a) A measure of *effect size often used in *meta-analysis. Symbolized Z_{FISHER}. (b) A *logarithmic *transformation of *Pearson's r correlation coefficients so that they can be used for *confidence limits and *significance tests. Also called *r-to-Z transformation. Sometimes abbreviated Fisher's Z_r.

Fishing Expedition   Any random looking around in the data gathered in a study to see if you can find some significant relationship. This is generally considered bad practice, especially if one is reporting the results of *significance tests. See *post hoc comparisons, *data mining. However, there are good ways to go "hunting" (versus bad ways to fish) for interesting relationships; see *exploratory data analysis, *data mining.

Fit   Refers to how closely observed *data match the relations specified in a *model, or how closely they correspond to an assumed distribution. See *goodness-of-fit.

Five-Number Summary   A way of describing a distribution of data associated with *exploratory data analysis. It is made up of the smallest value, the largest value, the median, the 25th percentile, and the 75th percentile. These are also the values used to construct a *box-and-whisker diagram.

Fixed-Effects Model   The typical *ANOVA design, in which the populations studied are (or are treated as) fixed categorical variables—users of Drugs A, B, and C, for example. The *independent variables and their *levels are determined (fixed) by the researcher. Also called "Model I ANOVA." See *random effects model, in which the researcher uses a random procedure to select the different levels of treatment.

Fixed Factor   Said of a variable in a *factorial *ANOVA design when the *levels of the variable are categorical or are categorized.

Floor Effect   A term used to describe a situation in which many subjects in a study measure at or near the possible lower limit (the "floor"). This makes analysis difficult since it reduces the amount of variation in the variable. Compare *ceiling effect.

For example, a study of counseling strategies to reduce suicide could be more difficult to conduct with subjects who were black women in their thirties, since this group has a very low suicide rate, one that could be hard to reduce further. It might be easier to conduct the study with white men in their sixties, since they have a much higher suicide rate.

Flow Chart (or Diagram)   A graphic illustration of progression through a system or the steps of a procedure. Used extensively in such fields as manufacturing and computer processing.

The following simple example from psychology briefly suggests the steps by which some environmental stimuli might be "processed" to eventually go into long-term memory.

Environmental stimuli → Initial perception → Short-term memory → Long-term memory

Figure F.1   Flow Chart

Flow Graph   Another term for *path diagram.

Fluctuations   In *time-series data, any short-term, back-and-forth movements that are unrelated to long-term *trends. See *spike, *moving average.

Sometimes fluctuations are so large it makes trends difficult to see. Global warming is an example. Even if the planet is gradually getting warmer, it still gets very cold in the winter (regular fluctuation); and sometimes the temperature can be quite low in the summer (random, nonpredictable fluctuation).

Focused Comparisons   Techniques of *meta-analysis for comparing *significance levels and *effect sizes by measuring the extent to which the studies' results are explained by an *independent variable.

Focus Group   A qualitative research tool pioneered in the 1940s and 1950s by Robert K. Merton, who called it the "focused group interview." The basic technique involves having about a dozen persons engage in an intensive discussion focused on a particular topic. It has been used extensively in market research among potential customers ("What would you think of this product?") and in planning political campaigns ("What do you see as this candidate's main weakness?"). It is increasingly used by survey researchers to help them design questionnaires.

Follow-Up Tests   Another term for *post hoc comparisons.

Forecasting   Predicting future events or quantities, such as the inflation rate next year, on the basis of information about past events or quantities. Forecasting is often contrasted with *estimation, by which one tries to determine the size of some existing quantity. Compare *prediction. See *ARIMA, *indicators.

   Researchers may forecast *events* (will an election take place?) or *event timing* (when will the election occur?) or *outcomes* (who will win the election?) or a *quantity* (what percentage of the electorate will vote for the Social Democrats?).

Formal Theory   Theory that uses symbols (logical or algebraic) rather than a natural language (like English or French) to state its propositions. Formal theorists manipulate formulas, not words. Their goal is to use symbols that will work in any context, to use forms that are not dependent upon content.

Formative Evaluation   *Evaluation research undertaken to find ways to improve, redesign, or fine-tune a program in its early stages. The focus is on the program's processes more than its outcomes, and the techniques are often *qualitative. Compare *summative evaluation.

Formula   A general *rule, principle, or statement of a relationship, usually expressed in mathematical or logical symbols. See *function.

FORTRAN   Short for FORmula TRANslator. One of the earliest *programming languages used for writing computer programs. Using commands such as "DO," "GO TO," and "READ," it "translates" English into a language a computer can use.

Forward Selection   A computer method for choosing which variables should be included, and in which order, in a *regression model or equation. Compare *backward elimination and *stepwise regression. Also called "step-up selection."

   For example, if you had 20 potential independent variables, the computer program would estimate 20 *simple regressions (one for each independent variable) and choose the "best" one, that is, the one that had the highest R^2 (explained the largest percent of the variance in the dependent variable). Then this variable would be tried in combination with the remaining 19 to find a second that (in a *multiple regression equation) produced the "best" pair (the pair with the largest R^2). Then it would use those two and search for a third, and so on until adding more variables no longer led to a significant increase in the R^2.

Fractal   A curve or a surface created by repeated subdivisions of the curve or surface; a geometric shape that is infinitely ragged, curvy, or otherwise irregular. Often any part of the fractal will contain a small-scale version of the whole. Fractals can be used to describe natural shapes such as crystals, clouds, coastlines, or snowflakes and have many applied uses such as weather forecasting and predicting patterns of population growth. See *chaos theory.

Fractile   Any division of a *distribution into equal units or fractions, such as fifths (*quintiles), tenths (*deciles), or hundredths (*percentiles). See *quantile.

Frame   See *sampling frame.

F Ratio (or Value or Statistic)   The ratio of explained to unexplained variance in an *analysis of variance, that is, the ratio of the *between-group variance to the *within-group variance. To interpret the F ratio, you need to consult a table of F values for a particular level of *statistical significance at the number of *degrees of freedom in your study. Named after Fisher, the inventor of analysis of variance. See *F test.

Frequency   The number of times a particular type of event occurs (such as the number of days it rained last year) or the number of individuals in a given category (such as the number of males under 21 years old who got speeding tickets this month).

Frequency Curve   (a) A smooth curve depicting the data in a *frequency polygon. (b) The curve representing a *probability density function.

Frequency Distribution   A tally of the number of times each score occurs in a group of scores. More formally, a way of presenting *data that shows the number of cases having each of the *attributes of a particular *variable.

 For example, the frequency distribution of final exam grades in a class of 50 students might be: 8 As, 20 Bs, 19 Cs, 1 D, and 2 Fs. In this example, the variable is final grade; the attributes are A, B, C, D, and F; and the frequencies are 8, 20, 19, 1, and 2. Compare *stem-and-leaf display.

 The following table presents a more elaborate frequency distribution of the data for these final exam scores. See the definitions of the various column heads (*class interval, *relative frequency, and so on) for more detail.

Table F.1   Frequency Distribution: Final Examination Grades in a Class of 50 Students

Final Grade	Class Interval	Frequency (f)	Relative Frequency	Cumulative Frequency	Cumulative Relative Frequency
A	90–99	8	.16	50	1.00
B	80–99	20	.40	42	.84
C	70–79	19	.38	22	.44
D	60–69	1	.02	3	.06
F	50–59	2	.04	2	.04

Frequency Function   See *probability density function.

Frequency Polygon   A line graph connecting the midpoints of the bars of a
*histogram.
      The following polygon depicts the same data as in the entry at
*histogram.

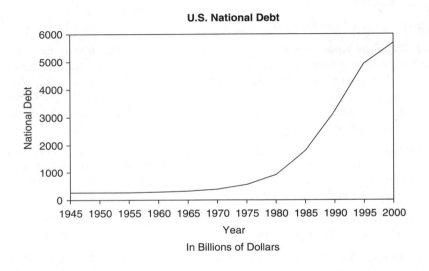

Figure F.2   Frequency Polygon

Frequentist   (a) An approach to studying probability that relies on the
frequency with which events have occurred in the past. (b) A person who
advocates the approach in definition (a). Frequentist methods are usually
contrasted with *Bayesian methods.

Frequentist Inference (or Probability)   The approach to statistics on which is
based the classical model of *significance tests, *hypothesis tests, and *confidence intervals. The frequentist approach is based on the assumption that a
statistic has been calculated an infinite number of times; see the discussion at
*sampling distribution. The main alternative is *Bayesian inference.

Friedman Test   A *nonparametric test of *statistical significance for use
with *ordinal data from *correlated groups. It is similar to the *Wilcoxon
test, but can be used with more than two groups. It is an extension of
the *sign test and is a nonparametric version of a *one-way, *repeated-
measures ANOVA.

F Scale   A widely used measure of the authoritarian personality (F for fascism) created in the aftermath of World War II by T. Adorno and colleagues. Not to be confused with the *F* ratio or *F* test.

F Statistic   See *F ratio, *F test.

F Test   A test of the results of a statistical analysis, perhaps most closely associated with, but by no means limited to, *analysis of variance (ANOVA). The *F* test yields an *F ratio or *F* statistic. This is a ratio of the *variance between groups (or explained variance) to the variance *within groups (or unexplained variance). To tell whether a particular *F* ratio is statistically significant, one consults an *F distribution table. The table is less and less frequently used, because computer programs routinely calculate the *F* ratio and report its level of significance. See *analysis of variance for an illustration. When the *F* test is applied to a regression equation, the *null hypothesis is that all the regression coefficients are zero.

Fully Recursive Model   Said of a *recursive *path analysis model when all the variables are connected by arrows, that is, when they are causally linked. Also called "just-identified model."

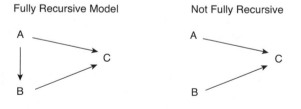

Figure F.3   Fully Recursive Model

Function   (a) A *variable that can be expressed in terms of another variable; also, a variable that varies with another variable. The term is often used loosely, if not very correctly, to mean a *cause. For example, the phrase "learning is a function of time spent studying" means that spending time studying causes learning. The functional relationship can be expressed in an equation: $L = f(T)$, where L stands for learning and T for time spent studying and f means "is a function of."

   (b) A mathematical expression of a relation between *independent variables and a *dependent variable; a formula or *equation such as a *regression equation.

(c) A *rule for turning each member of one *set into a member of another set. For example, $y = f(x) = 4x + 2$ says that to convert a number from set X into a number in set Y you multiply the number (x) times 4 and add 2 to it. Thus, the number 1 in set X would be 6 in set Y, 2 would be 10, 3 would be 14, and so on.

(d) In *set theory, a functional relation exists when each *element, x, in one set, X, is paired with an element y in set Y; this is written $y = f(x)$, meaning "y is a function of x." The set X is called the "domain" and the set Y is called the "range." A function can be plotted on a graph if for every value of x there is exactly one value of y. In the equation $y = f(x)$, Y is the *dependent variable and is plotted on the *y axis, and X is the *independent variable plotted on the *x axis.

Functionalism   Short for structural functionalism. A perspective on research in sociology and anthropology based on the assumption that social phenomena that are widespread and long lasting probably fulfill a social function. Hence, to explain a social form (structure) a functionalist will look for its usefulness for maintaining the society. Functionalism is an idea borrowed from anatomy, where the same assumption is made: Anatomical structures exist for a functional reason; the reason that, say, the kidney has the shape (structure) it does is that it could not function (as well) otherwise.

For example, a functionalist might study income inequality (the structure) in a society by looking for the ways it was functional for the society as a whole. Perhaps inequality increases motivation and thus stimulates the overall productivity and economic well-being of the society. Compare *conflict theory.

Functional Relationship   A relation that can be expressed in an *equation. See *function.

Funnel Plot   In *meta-analysis, a graphic method of detecting *publication bias. The *sample sizes are plotted against the *effect sizes of the studies. If publication bias is absent, the resulting graphic should be close to symmetrical and shaped roughly like a funnel or pyramid. See *file drawer problem.

Fuzzy Set   Said of *sets, or groups of variables, in which the extent of membership in the set is not precise. Since a set is a clearly defined group, a fuzzy set is a group the definition of which is allowed to vary or which can be hard to *operationalize. Based on complicated mathematics, it is sometimes used by analogy in the social sciences to discuss variables that are difficult to quantify, such as the degree of democracy in comparative studies of nations' political systems. See *complexity theory.

G   Sometimes short for *gamma.

G^2 Symbol for the *likelihood ratio test of *goodness-of-fit. The larger the sample, the more G^2 tends to approximate the *chi-squared distribution. The lower the G^2 for a particular independent variable, the more it contributes to the explanation of the dependent variable.

Gabriel's STP   One of several ways to adjust *significance levels in *multiple or *post hoc comparisons to reduce the chance of *Type I error. STP stands for simultaneous test procedure. See *Tukey's HSD test.

Gain Scores   See *difference scores.

Gambler's Fallacy   The mistake of treating *independent events as though they were *dependent.

   The familiar example has to do with tossing a fair coin. If after five consecutive heads you concluded that a sixth toss was more likely to come up tails, you would be committing the gambler's fallacy: You would be assuming that the sixth toss was dependent upon the previous five when, in fact, each toss is independent of the others.

   Or, suppose you shuffle an ordinary deck of cards and draw a card from it at random. The card is red (a heart or a diamond). You replace the card, reshuffle the deck, and draw a second card; this too is a red card. You repeat the process and draw a third red card. You commit the gambler's fallacy if you believe that, because you drew three red cards in a row, your fourth draw is more likely to be a black card. Because you replaced the card and reshuffled each time, each draw was an independent event, that is, it had no influence on subsequent events (and was not influenced by prior events). On the other hand, had you not replaced the red card each time, that would have made drawing a black card more likely. See *sampling with replacement.

G

The confusion arises in part from the fact that the best estimate, made *before* you begin, of your *probability of drawing four red cards in a row is quite low, .0625. But your best estimate of drawing four red cards in a row, *given that* you have already drawn three and replaced them, is .50. See *geometric distribution.

Game Theory A mathematical theory of competitive games in which each player wants to figure out the best way to play games given their rules. The games in question have to be strategic games, that is, not games of pure chance, but games based on knowledge, including knowledge of what the other players are likely to know. Game theory has been widely applied as a model of human decision making and action in such fields as economics and military strategy. See *maximin strategy, *minimax strategy, *zero-sum game.

Gamma [Γ, γ] (a) Sometimes called "Goodman and Kruskal's gamma." A *measure of association for *ordinal variables. It is a *symmetric, *PRE statistic. Gammas range from −1.0 to +1.0. When calculating gamma, the ranks of the ordered categories are not treated as interval scales. Compare *Spearman's rho. (b) *Euler's constant. (c) A *regression coefficient in *hierarchical linear modeling.

For example (definition a), if knowing how a sample of citizens ranked on one variable, such as opinion about gun control (strongly opposed, opposed, in favor, strongly in favor) always enabled you to predict how they would rank on attitude toward the death penalty (strongly opposed, opposed, and so on), the gamma indicating the association between those two variables would be −1.0. But, since, in fact, there are people who oppose both gun control *and* the death penalty, the gamma might be something more like −.8.

Gantt Chart A type of bar chart used to plan projects, especially to schedule tasks and keep track of whether they are being completed on time.

Gap Statistic A statistic for estimating the number of clusters in a *cluster analysis.

Garbage In, Garbage Out See GIGO.

GARCH Generalized Autoregressive Conditional Heteroscedasticity. GARCH models are used to describe and predict the volatility of financial time-series data. See *ARIMA.

Gaussian Distribution Another term for *normal distribution.

Gauss-Markov Theorem A theorem which states that if the *error terms in a multiple regression are uncorrelated and have equal variances, then the *least squares criterion for calculating the *coefficients is better than any other estimator.

Geisser-Greenhouse *F* Test   An **F* test that adjusts the critical value upward in a *repeated-measures ANOVA to correct for violations of the *sphericity assumption. Sometimes called the Geisser-Greenhouse conservative test.

Geisteswissenschaften   German for "human sciences," (literally, "spiritual" or "mental" sciences) including the social and behavioral sciences as well as the humanities. The term has mostly been used in contrast with the natural sciences (*Naturwissenschaften*). Classic debates about research methods in Germany in the 19th and 20th centuries centered around the question of whether these two broad categories of science can or should use the same or different methods. Advocates of the distinctiveness of the human sciences were among the most important early critics of *positivism. See **Verstehen.*

General Effect   The impact of a *variable on another that does not take into account differences among groups. See *effect. Compare *conditional effect, *interaction effect.

   For example, graduating from college improves average earnings for all (general effect), but it does so more for some groups (conditional effect).

Generalizability   The extent to which you can come to conclusions about one thing (often, a *population) based on information about another (often, a *sample). Compare *external validity, *inferential statistics.

   For example, when a national burger restaurant wants to see whether a new sandwich will sell, it promotes the sandwich in a few communities that are assumed to be representative of the nation. If that assumption is correct, the company can accurately generalize from the sales figures in the handful of communities to how the new product will sell in the rest of the country.

Generalizability Theory   An alternate way to estimate *reliability suggested by Cronbach; it identifies the different sources of *error in a measure rather than simply estimating the total error. The procedures for doing this are complicated, which may be why most researchers continue to estimate reliability using simpler statistics, such as *Cronbach's (same statistician) alpha.

Generalization   (a) A statement covering all the members of some group or class. (b) The process of coming to such a conclusion by *inference; the inference is usually *inductive. See *external validity.

   Few issues are as central or as controversial in the social and behavioral sciences as the characteristics of (true or probable) generalizations and the best methods of arriving at them. See *theory.

General Linear Model   A common set of statistical *assumptions upon which are based *regression, *correlation, and *analysis of variance—in

G

short, the full range of methods used to study the *linear relations between one continuous dependent variable and one or more independent variables, whether continuous or categorical. The model is "general" in that the kind of independent variable is not specified.

The basic idea is that the relation between a dependent variable and the independent variables can be expressed as a linear equation containing a *term for the weighted sum of the values of the independent variables—plus a term for everything that we do not know about, which is called an *error term. The method used to decide how much weight to give to the independent variable(s) is the *least squares criterion. Compare *polynomial regression analysis.

You can get a good practical "feel" for the general linear model by spending a few hours calculating by hand standard deviations, Pearson correlations, regression equations, t-tests, and F tests. You will quickly discover that many of the steps in these calculations are generally the same.

Generalized Linear Models   Statistical techniques used when the assumptions of the *general linear model are not met, as when the dependent variable is not continuous and not distributed normally. These nongeneral features of the dependent variable are general*ized,* usually by logistic *transformations. In this context, transformations of the dependent variable are called *link functions. *Logistic regression and *log-linear models are examples, as are *probit models.

Generalized Least Squares   A means of calculation used (instead of *ordinary least squares) in *regression analysis when there is a nonrandom pattern to the *error terms (that is, *heteroscedasticity in residual error terms) or when *autocorrelation or serial correlation biases the results.

Generalized Maximum Likelihood Ratio Test   A statistic formed by taking the ratio of the *maximum likelihood of drawing a particular *sample if a particular hypothesis were true to the maximum likelihood of drawing that sample if that hypothesis were not true.

General Social Survey (GSS)   An annual or biannual survey of a representative sample of (about 1,500) American adults conducted by the National Opinion Research Center. Respondents are asked several questions about their backgrounds and for their opinions about many social and political issues. The results of these surveys are available to, and are widely used by, other researchers for *secondary analyses.

Generation Effects   The effects on individuals of growing up in or being members of the same generation or age group. See *cohort effects.

GenStat   Short for GENeral STATistics, a statistical package used especially by researchers working with biological and environmental statistics.

Geometric Distribution   A *probability distribution of the number of *failures before the first *success in a series of trials with only two possible outcomes (*Bernoulli trials). See *Pascal distribution.

  In the following example, the table gives the probability of getting tails on the first, second, and so on, flip of a coin and of getting a heart on the first, second, and so on, draw (with *replacement) from an ordinary deck of cards. Consulting the table, you can conclude, for instance, that only about 3% of the time would you need five flips of a coin to get tails. Only about 8% of the time would you need to draw five cards before getting a heart. In these two examples, as with all geometric distributions, the most probable number of trials (before obtaining the first success) is always one trial.

Table G.1   Geometric Distribution. Probability of Getting the First Success on Each Trial

Trial	Tails	Heart
1	.50	.25
2	.25	.1875
3	.125	.1406
4	.0625	.1055
5	.03125	.0791

Geometric Mean   See *mean.

Giga   A prefix indicating a multiple of 10,000,000,000; thus a gigabyte is 10,000,000,000 bytes.

GIGO   Short for "garbage in, garbage out." A brutal way of putting an undeniable principle: No matter how good your computer and your statistical analysis package, if you put poor *data (data that are neither *reliable nor *valid) into your computer, it will give you poor results.

Gini Index (or Coefficient)   A measure of inequality or dispersion in a group of values, such as income inequality in a population. It ranges from 0 to 1, with larger coefficients indicating greater dispersion or inequality. It is calculated by taking the mean difference between all pairs of values and dividing the result by 2 times the population mean. The Gini statistic is derived from the *Lorenz curve. Sometimes called the "coefficient of concentration" or "Gini's ratio."

Glejser Test   A test statistic used to diagnose the problem of *heteroscedasticity in the *residuals of a *regression equation.

GLIM   Short for Generalized Linear Interactive Modeling, a *statistical package.

G

GLM   Abbreviation for the *general linear model and for *generalized linear models.

Global   Sometimes used to mean general or pertaining to all variables, but not necessarily having anything to do with a sphere or worldwide phenomena.

GNI   *Gross national income.

GNP   *Gross national product.

Goedel's Proof   The demonstration by Kurt Goedel (in 1931) that a formal system, such as logic or mathematics, cannot prove its own basic axioms. This has sometimes been taken more generally to indicate that we can have little certainty in our claims to knowledge. Compare *Heisenberg's uncertainty principle.

Going Native   In anthropological research, especially *participant observation, said of investigators who lose their *objectivity and essentially become a member of the culture they are studying. Participant observation requires that researchers become members of the groups they are studying—at least to *some* extent. Controversy and difficult judgments occur when trying to specify to *what* extent.

Goldfeld-Quandt Test   One of several statistical tests to detect *heteroscedasticity in the *residuals of a *regression equation.

Gold Standard   Used loosely to mean the best, state-of-the-art methods for research problems. Perhaps most often used to refer to *randomized clinical trials. Ironically, in monetary policy, where the term originated, the gold standard has long been abandoned as ineffective.

Gompertz Curves   Growth curves that are shaped like an elongated letter S. The curve describes a growth rate that is initially small, increases for a time, and then levels off. Compare *J curve.

Goodman and Kruskal's Gamma   See *gamma.

Goodman and Kruskal's Lambda   See *lambda, definition (c).

Goodman and Kruskal's Tau   See *tau.

Goodness-of-Fit   How well a *model, a theoretical distribution, or an *equation matches actual *data. See *goodness-of-fit test.

   For example, in *regression analysis, the question is: How closely does the *regression line (formed of the *predicted scores) come to summarizing the observed scores? The coefficient of *determination, or R^2, is a measure of the goodness-of-fit for a regression line. In the following simple graph, line A fits the data better ("gooder") than line B.

Graph   135

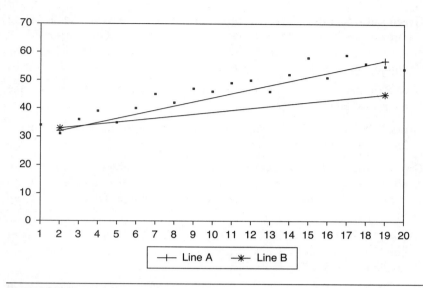

Figure G.1   Goodness-of-Fit

Goodness-of-Fit Test   A statistical test to see whether a model fits a set of data, whether it matches a theoretical expectation. The most commonly used test is the *chi-squared test; the bigger the chi-squared statistic, the poorer the fit, the smaller the better. That is because a small chi-squared statistic indicates a small difference between the model and the data. The *likelihood ratio test is also used as a measure of goodness-of-fit. See also *Kolmogorov-Smirnoff test.

For example, say a researcher wants to know if bankruptcies are randomly distributed throughout the year. The theoretical distribution in this case would be equal probability; the *expected frequency would be that 1/12 of the bankruptcies occur in each month (with slight adjustments for the longer and shorter months).

Granger Causality   Techniques in *regression analysis of *time-series data in economics to investigate whether variables that are statistically associated are causally related. The method involves taking into account the fact that causes do not operate instantaneously; they take time; and these time lags are incorporated into the model. See *cause, *confounded, *lurking variable.

Graph   A diagram showing the values of a *variable or showing a relationship between two variables. For examples see *bar graph, *curvilinear relation, *histogram, *frequency polygon, and *scatterplot.

G

Greco-Latin Square An extension of the *Latin square method of allocating *treatments in a *within-subjects *factorial experiment. The extension is done by adding a different Greek letter to the Latin letter in each cell. As with the Latin square method, the goal is *counterbalancing *order effects.

Grid Sampling A form of *cluster sampling in which the clusters are the areas marked off by a grid.

Gross The amount before any deductions. For example, gross income is income before deducting expenses. Compare *net.

Gross National Income (GNI) The total of a nation's earnings from all sources in a year. It is a common method of calculating the same information as the *gross national product. By definition, GNI and GNP are identical.

Gross National Product (GNP) The total value of a nation's output of goods and services in a year. It is a frequent *indicator of a nation's economic health. See *gross national income.

Grounded Theory A method for constructing theory; the theory grows out of extensive direct observation and inductive methods in a *natural or *non-experimental setting. It very often involves the analysis of texts. Grounded theory is frequently a goal of *ethnographic research. The term is used loosely to mean any theory based on data and very specifically to mean methods based on the work of Anselm Strauss. Both the method and the conclusions reached by using it are labeled grounded theory—"grounded theory was used to produce this grounded theory." See *a posteriori, *inductive.

Grouped Data Data recorded as numbers of cases in *class intervals. Compare *raw data, *aggregate data.

Group Effect The influence on individuals of being members of one group rather than another, such as the effect on one's earnings of being female. See *contextual effects, *t-test, *analysis of variance.

Grouping Another term for *collapsing.

Group Randomized Trial A method in which groups, not individuals, are assigned randomly to treatment or control conditions. All individuals in each group receive the same treatment. Compare *quasi-experiment, *cluster randomizing.

Growth Curve Analysis Statistical methods for studying development over time, originally for uncovering trajectories in the biological development of individuals. Also called "latent curve models" and "latent trajectory models." The models are developed using either *hierarchical linear modeling or *structural equation modeling. See *time series, *longitudinal studies.

GSS   *General Social Survey.

Guttman Scaling   A method of scale construction created by Louis
Guttman. It was originally designed to be used after the data were collected
to see if the items in an *index could be arranged as a *scale, that is, in the
order of the strength of the items. See *Bogardus social distance scale.
    For example, national surveys often ask questions about abortion
roughly as follows:

Do you favor a woman's right to have an abortion if:

1. Having the baby would threaten her life?

2. The fetus is deformed?

3. She is too poor to care for the child properly?

4. She does not want any more children?

If these items form a Guttman scale, the vast majority of people who answer
the questions will do so in a scalar pattern: People who say yes to question
number 4 will also say yes to questions 3, 2, and 1; those who say yes to number
3 will also say yes to 2 and 1, but not necessarily to 4, and so on. If these items
do not form a scale, there will be no pattern to the answers: People who say yes
to number 4 will be as likely as not to say no to numbers 3, 2, and 1.

H   The usual symbol for the statistic computed for the *Kruskal-Wallis test of statistical significance.

H₀   Symbol for the *null hypothesis.

H₁   A symbol for an *alternative or *scientific hypothesis.

h²   Symbol for the *communality of an item (or indicator or variable) in a *factor analysis.

Hₐ   A symbol for the *alternative or *scientific hypothesis.

Halo Bias (or Effect)   A tendency of judges to overrate a performance because the subject did well in an earlier rating or when rated in a different area. Also used to mean the tendency to make more favorable judgments about physically attractive people, an effect which has been studied in areas such as job performance ratings and sentencing of convicted criminals.

     For example, say a student has taken two courses from a professor and has done exceptionally well in each. In a third course, she writes an ordinary paper. This paper might receive a higher grade than it deserves because the student's earlier good work creates a halo effect.

Hanning   A technique for *smoothing data in a *trend line. Compare *running medians, *moving average.

Hard Sciences   Natural sciences, especially physical sciences. Often contrasted with soft sciences such as psychology, sociology, political science, and economics. Some scholars think this distinction indicates a real difference; others dismiss it as mere "physics envy." Among the social and behavioral sciences, economics and psychology more often make successful claims to "hardness," largely because they have used quantitative research methods longer than the others.

H

Hardware In computer jargon, the physical components of a computer; the machine without any operating instructions or *software.

Harmonic Mean A *measure of central tendency used mainly in comparing average rates. See *mean for details.

Hash Short fluctuations in *time-series data.

Hawthorne Effect A tendency for subjects of research to change their behavior simply because they are being studied. So called because the classic study in which this behavior was discovered was in the Hawthorne Plant in Illinois. In this study, workers improved their output regardless of the changes in their working conditions. Compare *John Henry Effect. See *artifact.

Hazard (a) Sometimes use to mean chance or *probability. (b) A cause of possible loss. The word comes from Arabic for a dice game. Compare *survival analysis.

Hazard Function A measure of how likely an individual is to experience an event (death, illness, etc.) given the age of the individual, and given that the individual has not yet experienced the event. Originally developed in medical research, this measure of *risk is used more broadly now.

Hazard Rate An estimate of the probability of failure of a system or of a component over time. Also called hazard ratio. See *hazard function.

Heisenberg Uncertainty Principle Because we cannot study the atom without affecting it, we cannot know (we must be uncertain about) what it might be like without our interference, when we are not studying it. This principle has been extended by some writers to areas and methods of research other than atomic physics, such as *participant observation. Compare *Goedel's proof.

Helmert Contrast A technique in *analysis of variance in which each level of a *factor (*variable) is tested against the mean of the remaining levels.

Helsinki Declaration Ethical principles for research, especially research involving human subjects. An extension of principles first importantly codified in the *Nuremberg code.

Herfindahl Index A measure of *concentration, usually of firms in a sector of the economy. Compare *Gini index.

Hermeneutics A philosophical theory of interpretation, originally of written texts, especially of biblical passages, but now used more broadly. Most importantly developed and used in Germany. See *Verstehen.

Heterogeneous Generally, mixed or diverse. Used to describe *samples and *populations with high *variability. See *variance, *standard deviation.

Heteroscedasticity   A situation in which there are considerably unequal
*variances in the *dependent variable for the same values of the *indepen-
dent variable in the different populations being sampled and compared in a
*regression analysis or an *ANOVA. Also in one population, when the
error variance is not constant for all levels of the independent variable.
Comes from *hetero,* meaning other or different and *scedasticity,* meaning
tendency to scatter. Heteroscedasticity violates the *assumption of
*homoscedasticity or *equality of variances. This violation, if serious
enough, compels the use of *nonparametric statistics. Although there are
statistical tests to detect heteroscedasticity, it is common to check for it
using scatter diagrams, such as the following.

In the scatter diagram, the variances in the dependent variable (salary)
are very similar at low levels of the independent variable (years of experience), but much less so at its upper levels.

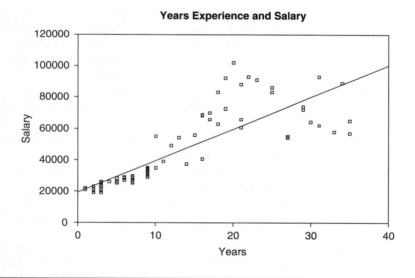

Figure H.1   Heteroscedasticity

Heuristic   (a) Generally, instructive or pedagogical. (b) Having to do with
methods that aid learning by exploratory, or trial-and-error methods; said of
a *computer program that can learn from its mistakes (by eliminating trials
that do not work) and/or that can teach people how to use the program by
learning from their mistakes. (c) *noun* A rule of thumb, a procedure for
making decisions. Compare *algorithm.

Heuristic Assumption   An *assumption made more because it is useful for
teaching or research purposes than because it is believed to be true.

H

Hexadecimal Notation A number system that uses the ten numbers 0 through 9 and the six letters A through F. It is useful in computer programming.

Hierarchical Models Two or more models arranged so that each higher model contains all the components of the next lower model plus at least one additional component. See *multilevel models.

Hierarchical Linear Models (HLM) An alternative to *OLS regression models used when data are found in *nested categories or levels, such as college students taking Sociology 201, sociology majors, majors in any of the social sciences, undergraduates in the university, or undergraduates in the state. The method's advantage is that it makes it possible to separate the variance into components explaining the effects of different levels of analysis, such as the effects of a course, of a major, or of being undergraduates. Also called "covariance components models," "multilevel modeling," "Bayesian linear models," and, mostly by economists, "random coefficient models."

Hierarchical Regression Analysis (a) A method of regression analysis in which *independent variables are entered into the regression equation in a sequence specified by the researcher in advance. The hierarchy (order of the variables) is determined by the researcher's theoretical understanding of the relations among the variables. Hierarchical techniques are often contrasted with *stepwise regression, in which the order of the variables is determined by a computer program using statistical associations among the variables in the particular data set.

It is common to combine hierarchical and stepwise procedures. For example, the researcher will enter the variables in groups or "blocks." The order of the blocks is determined by the researcher as in hierarchical regression, but within each block the order is determined by a computer program using stepwise techniques (*backward elimination and *forward selection). In part because of such possible combinations, usage tends to vary quite a bit.

(b) A type of *regression model which assumes that when a higher order *interaction term is included, all the lower order terms (*main effects) are also included.

Higher Order ANOVA An ANOVA with three or more *independent variables.

Higher Order Interaction Effects Interaction effects between more than two *independent variables. See *first-order interaction effect.

Whenever a *factorial model has more than two *factors or independent variables, higher order interactions are possible, and the number of possible interactions increases rapidly as the number of factors increases.

Higher Order Partials (or Correlations) Fully, "higher order partial correlations." Correlations that *control for more than one *variable in a complex, *multivariate relationship. A *zero-order correlation controls for no other variables; a *first-order correlation controls for one, a second-order correlation for two, and so on. All beyond the first order are higher order. Compare *partial relation.

For example, when computing a correlation between persons' education levels and their incomes, a researcher might wish to control for other variables that could influence the relationship between education and income. If three variables were controlled (such as age, sex, and ethnicity), the correlation would be a higher order (third-order) partial correlation.

High-Level Language   *Software for programming computers. The programmer writes programs in the high-level language, which then "translates" the program into language that the machine (*hardware) can understand. Examples include BASIC, C, COBOL, and FORTRAN.

Highspread   The range of values between the *median and the highest value in a *distribution. Compare *lowspread.

Hinge   The point in a *distribution that divides the scores at the 1/4–3/4 mark. The lower hinge is the score at the 25th *percentile, that is, the point in a distribution above which 3/4 of the scores are located. The upper hinge (at the 75th percentile) is the point with 1/4 of the scores above it and 3/4 below it. See *exploratory data analysis, *box-and-whisker diagram, *five-number summary.

Histogram   A *bar graph for *variables measured at the *interval and *ratio levels. Because the data in a histogram are interval or ratio, the bars should touch; in a bar graph showing *nominal or *ordinal data, they should not. The data shown here are the same as those used to construct the *frequency polygon.

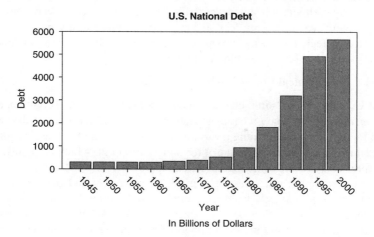

Figure H.2   Histogram

Historical Sciences   Disciplines in which *cause and effect are separated by relatively long periods—including, but not limited to, topics taught in traditional history courses.

For example, geology is a historical science; the formation of continents takes millions of years. On the other hand, while chemical reactions occur over time, the amounts of time are usually very small by human standards, so chemistry is not considered a historical science. In psychology, personality development, which takes years, could be thought of as a historical science; perception, which takes milliseconds, would not be. Compare *longitudinal study.

Historicism   The original form of *cultural relativism stressing that different eras can only be understood in their own terms. Like other forms of relativism, it can lead its adherents to doubt all claims to knowledge. Compare *anachronism.

History Effect   An event that intervenes in the course of one's research and makes it difficult if not impossible to interpret the relations among *independent and *dependent variables. Usually listed among *threats to validity. See *extraneous variable, *confounding variable.

Suppose a city government begins an experiment with sensitivity training (*independent variable) to improve intergroup relations among its employees. Attitudes are measured before the training begins and are to be measured again after 10 weeks. In the ninth week a dramatic event occurs that could influence subjects' attitudes (such as a race riot or a controversial affirmative action ruling). Any differences in attitudes measured after the 10th week of training could be due to the training; or the training effects might have been affected by the event (the history); there is no way to tell.

History effects are not necessarily dramatic and are perhaps more of a problem when they are less striking and can more easily go unnoticed. History effects are always a potential threat to validity for any nonlaboratory study that lasts more than a few hours.

HLM   *Hierarchical Linear Modeling or Models.

Hold Constant   To "subtract" the effects of a *variable from a complex relationship so as to study what that relationship would be if the variable were in fact a *constant. Holding a variable constant essentially means assigning it an average value. See *control for and *partial out, which are equivalent expressions for the same mathematical operations.

For example, in a study of managerial behaviors and their effects on workers' productivity, a researcher might want to hold the education of the managers constant. This would especially be the case if she had reason to believe that different kinds or amounts of education might lead managers to behave differently.

H

Holism An *assumption that groups, collectivities, or wholes can be more than, or different from, the sum of their individual parts. This leads to an approach to research that stresses studying wholes or complete systems, rather than analyzing individual parts. Compare *reductionism, *methodological individualism.

For example, a holist might say that an organization, such as General Motors, exists independently of the individuals who work for or own it. After all the individuals who today own or work for General Motors quit, retire, die, or sell their stock, the corporation could still exist—as long as those people were replaced by others. Since General Motors is not just the sum of its parts, it makes sense, holists would say, to talk of GM "wanting," "planning," "deciding," and so on. Such desires, plans, and decisions are not reducible merely to those of individuals.

Holms Procedure A modification of the *Bonferroni adjustment technique for multiple statistical tests. It involves using a range of *p values with smaller p values required for the larger statistical associations and larger values allowed for the weaker associations.

Homogeneity of Variances An assumption that populations from which samples have been drawn have equal variances. If this assumption is not true, *test statistics may be inaccurate. Also called *equality of variance assumption. See *homoscedasticity and, for an illustration, *heteroscedasticity.

Homogeneous Generally, the same or similar. Used to refer to *populations and *samples that have low *variability. Compare *heterogeneous.

Homoscedasticity *Homogeneity of variances. A condition of substantially equal *variances in the *dependent variable for the same values of the *independent variable in the different populations being sampled and compared in a *regression analysis or an *ANOVA. Comes from *homo*, meaning the same or equal and *scedasticity*, meaning tendency to scatter (skedaddle?). *Parametric statistical tests usually assume *homoscedasticity. If that assumption is violated, results of those tests will be of doubtful validity. See *heteroscedasticity for a graphic representation.

Honestly Significant Difference (HSD) Test See *Tukey's Honestly Significant Difference Test.

Hotelling's t^2 Test An extension of the *t-test to *multivariate research problems. Also called Hotelling's T^2 test.

HRAF *Human Relations Area Files.

HSD Test See *Tukey's Honestly Significant Difference Test.

Human Capital A kind of *capital (resources that can produce income) that exists within persons rather than external to them. Knowledge, skill, and strength are examples of human capital. Economists often think of education as an investment in human capital. Compare *cultural capital.

Human Relations Area Files (HRAF)   A collection of anthropological information about hundreds of human cultures throughout the world divided into several hundred categories of information. The HRAF is widely used by researchers doing *secondary analyses of data on cross-cultural topics.

Hypergeometric Distribution   A *probability distribution used for studying *sampling without replacement, that is, when each selection (or trial) changes the probability of the outcome of the next. Each member of the population can be categorized into one of two classes, such as success/failure or yes/no.

Hyperplane   A plane of three or more dimensions. While hard to picture, the hyperplane is a widely used concept in *multiple regression. See *regression plane for an illustration.

Hypothesis   A tentative answer to a research question; a statement of (or conjecture about) the relationships among the *variables that a researcher intends to study. Hypotheses are sometimes testable statements of relations. In such cases, they are usually thought of as predictions, which if confirmed, will support a *theory. See *alternative hypothesis, *null hypothesis, *research question.

For example, suppose a social psychologist theorized that racial prejudice is due to ignorance. Hypotheses for testing the theory might be as follows: If (1) education reduces ignorance, then (2) the more highly educated people are, the less likely they are to be prejudiced. If an attitude survey showed that there was indeed an *inverse relation between education and prejudice levels, this would tend to confirm the theory that prejudice is a *function of ignorance.

Hypothesis Testing   The classical approach to assessing the *statistical significance of findings. Basically it involves comparing empirically observed *sample findings with theoretically expected findings—expected if the *null hypothesis is true. This comparison allows one to compute the *probability that the observed outcomes could have been due to chance or *random error. See *alpha error, *beta error.

For example, suppose you wanted to study the effects on performance of working in groups as compared to working alone. You get 80 students to volunteer for your study. You assign them randomly into two categories: those who work in teams of four students and those who would work individually. You provide subjects with a large number of math problems to solve and record the number of answers they got right in 20 minutes. Your *alternative or *research hypothesis might be that people who work in teams are more efficient than those who work individually. To examine the research hypothesis, you would try to find evidence that would allow you to reject your null hypothesis—which would probably be something like: There is no difference between the average score of students who work individually and those who work in teams.

The outcomes of a decision in hypothesis testing are often depicted in a matrix as follows. (Compare the similar matrix illustrating *signal detection theory.)

Table H.1   Possible Outcomes of a Hypothesis Test

		Decision	
		Null Hypothesis Is True	*Alternative Hypothesis Is True*
Reality	*Null Hypothesis Is True*	Correct retention	Type I (alpha) error, wrong rejection
	Alternative Hypothesis Is True	Type II (beta) error, wrong retention	Correct rejection

Hypothetico-Deductive Method   A name philosophers of science sometimes use to describe the general logic of research that uses *hypothesis testing to draw its conclusions. The basic steps are: using *theory to deduce testable hypotheses; gathering data and using statistical inference to test the hypotheses; supporting or challenging the theory based on the outcome of the hypothesis testing.

ICC   *Intraclass Correlation.

Ideal Index   An index computed by taking the *geometric mean of two indexes. The goal is to have the *biases in the two indexes offset one another. Sometimes called Fisher's ideal index.

Idealism   A wide range of philosophical doctrines and perspectives method-ologically important for their belief that the mind and its ideas (not exter-nal experience) are the ultimate source and criteria of knowledge. Compare *empiricism, *epistemology.

Ideal Type   A term introduced by Max Weber to refer to a *model or a pure conceptual type. "Ideal" does not refer to the best or most desirable; rather, it means "pertaining to an idea." In modern English usage, "conceptual type" would capture Weber's meaning. See *concept, *construct.
  For example, one could describe an ideal-typical bureaucracy; this would be as conceptually pure a bureaucracy as one could imagine, that is, an organization that worked only according to bureaucratic principles. An ideal type is used as a category or a concept to guide research. Thus, if you defined how a pure (ideal type) bureaucracy would work, you could use this as a standard to hold up to actual organizations to see how bureaucratic they were. Without such a standard it would be difficult to say whether one organization was more bureaucratic than another.

Identification Problem   An analytic difficulty that arises in *regression analysis when one has more unknowns than can be independently estimated from the available data; in other words, it occurs when there are too many unknowns in a causal *model for a solution to be possible.

Identity Matrix   A matrix in which all the elements on the diagonal are 1.0 and all the other elements are zero. See *matrix algebra.

149

Ideology   A system of beliefs held by a group that tends to serve the interests of that group. In research reports, one generally reserves the term "ideology" for positions one really does not like at all.

Idiographic   Used to describe research that deals with the individual, singular, unique, or concrete. Idiographic is often contrasted with *nomothetic. Compare *generalizability.

i.i.d.   Abbreviation for "independent and identically distributed." For example, if values for three *predictor variables were sampled from a single *population and they were all *independent and *normally distributed, they would meet the criteria for i.i.d. Compare *correlated groups design.

Illusory Correlation   See *spurious correlation. Also called nonsense correlation.

Impact Assessment   A variety of evaluation research that focuses on the long-term effects (planned and unplanned) of a program or intervention. It is sometimes distinguished from an outcomes assessment in which short-term and intended consequences are examined.

Implicit Measures   Indirect measures of psychological states, traits, and conditions. They are used as indicators of variables that might not be accurately measured more directly. The kind of physiological measures used by lie detectors are one example. Another category clocks the amount of time it takes subjects to answer questions about different topics; response time is taken as an implicit measure of attitudes about those topics.

Imputation   Methods of replacing missing values in a data set. The simplest is *mean substitution, in which the missing value for a case is replaced with the mean score of the other cases. For example, if teachers' income were a variable in your study and you were missing the incomes for a few teachers, you could use the mean for the other teachers as an estimate of the missing values. Most methods of imputation are controversial; many experts recommend deleting cases with missing values and recommend against any imputation. See *listwise deletion, *missing values procedures, *extrapolation, *interpolation.

"Hot deck" imputation and "cold deck" imputation are techniques that are still used; they were named when computerized data sets were stored on decks of computer punch cards.

Incidence Rate   The number of times something occurs divided by the size of the population in which it could occur. It is the *relative frequency with which new cases of something took place—crime, marriage, disease, and so on.

Increment   A small change in the value of a variable—often a positive change. A negative change is sometimes called a decrement.

Independence   (a) In *probability theory, a state in which the occurrence of one event does not change the probability of another event, that is, when one event does not depend on another. The numbers in a table of random numbers are independent. (b) In statistics and research design, two variables are independent when the value of one has no effect on the value of the other. (c) More generally and loosely, two variables are said to be independent when measures of association between two variables are small and not *significant; however, statistically associated variables may be independent—such as a person's age and the price of cheese over the past 20 years. Compare *gambler's fallacy, *sampling with replacement, *orthogonal.

Independent Event   In probability theory, said of an occurrence that is not *conditional upon or conditioned by another. See *independence, *dependent event.

Independent Samples (or Groups)   Groups or samples that are unrelated to one another, that is, when the measurements of subjects in one group have nothing to do with the measurements of subjects in the other group.

   For example, individuals put into different treatment groups by *random assignment would be independent. By contrast, subjects assigned to treatment groups according to the time of day they arrived at a clinic would not be independent. Early risers might have traits in common, as might people who arrived in the evening (they might work during the day). Also, students sampled from the same school would not be independent samples of all students because students in the same school would be likely to have some things in common. Compare *between-subjects variable, *correlated groups design.

Independent Variable   The presumed *cause in a study. Also a variable that can be used to predict or explain the values of another variable. A variable manipulated by an experimenter who predicts that the manipulation will have an effect on another variable (the *dependent variable).

   Some authors use the term "independent variable" for experimental research only. For these authors the key criterion is whether the researcher can manipulate the variable; for nonexperimental research these authors use the term *predictor variable or *explanatory variable. However, most writers use "independent variable" when they mean any causal variable, whether in experimental or nonexperimental research. Some even use it in pure *forecasting, where no causal connection is implied, as when variations in the starting date of the migrating season are used to predict the severity of winter temperatures.

Index   (a) Any observable phenomenon that is used to indicate the presence of another phenomenon, as when attending church is used to indicate religious commitment. Compare *proxy variable, *indicator. (b) A number, often a *ratio, meant to express simply a relationship between two variables or between two measures of the same variable. (c) A group of individual measures which, when combined, are meant to indicate some more general characteristic. Compare *scale.

For example (definition b), indexes measuring access to medical school for various groups could be calculated by dividing a group's percentage of students in medical schools by its share of the population of medical school age. If, say, women made up 50% of the 21- to 25-year-olds and were 40% of the medical students, their access index would be .8 (40/50 = .8). Or, if white males were 40% of the 21- to 25-year-old population and were 56% of the medical students, their index would be 1.4 (56/40 = 1.4).

For example (definition c), in survey research, political tolerance might be measured by an index composed of six questions about whether the respondent favored such things as free speech for religious outsiders, the right to demonstrate for political radicals, and so on. Scores on the index could range from 0 (for those answering none of the questions in the tolerant way) to 6 (for those answering all of the questions in the tolerant way). This kind of index is used to measure an *ordinal variable.

Indexization Adjusting wages, taxes, Social Security benefits, and so on to an *index number, such as the *consumer price index.

Index Number A measure of change as compared to a particular *base year or date in a *time series. The base year is often set at 100. Used most often with economic variables. See *consumer price index, *Laspeyres's index.

Indicator (a) A specification of how we will recognize or measure a *concept, as an IQ test could be taken as an indicator of the concept intelligence. (b) A variable that can be used to study another variable because it affects or is correlated with it, as absenteeism could be an indicator of employee satisfaction. Clusters of related indicators are *dimensions. See *indicator variable. See *social indicators.

Indicator Variable (a) Another term for *manifest variable, that is, an observable variable one uses to study a *latent (unobservable) variable. (b) Another term for *dummy variable.

Indices An alternative way of pluralizing *index, and one that is more grammatically correct (or traditional) than "indexes."

Indirect Proof Demonstrating the truth of a conclusion by showing that its logical opposite is self-contradictory or contradicts known truths.

Individual-Difference Variables Another term for *background variables.

Induction The logical process of moving from particular information to general conclusions. This would include using statistical methods to form generalizations by finding similarities among a large number of cases. The generalizations derived in this way are probabilistic. For example, if 90%

of the members of the U.S. Congress were lawyers, the chances of any individual member of the Congress being a lawyer would be 9 out of 10.

Inductive   Said of research procedures and methods of reasoning that begin with (or put most emphasis on) observation and then move from observation of particulars to the development of general *hypotheses. Often used to describe *ethnographic research. Compare *empirical, *deductive.

Inductive Statistics   Another term for *inferential statistics.

Inference   The act of using one statement (or judgment or proposition or generalization) to derive a new one. The truth of the new statement is held to follow logically (*deductively or *inductively) from the original statement.

Inferential Statistics   Statistics that allow one to draw conclusions or inferences from data. Usually this means coming to conclusions (such as estimates, generalizations, decisions, or predictions) about a *population on the basis of data describing a *sample. See *statistical inference.

Influential Observation (or Case)   A piece of data that would, if removed from the other data, greatly change conclusions drawn from the data, especially in *regression analysis. Often an *outlier.

Influence Statistics   In *regression analysis, statistics designed to assess the effect (influence) of particular observations on the *regression coefficients. See *regression diagnostics.

Informant   A person who provides a researcher, often an anthropologist doing *field research, with information. Informants are sometimes distinguished from "mere" *subjects or *respondents by the extent of their personal interaction with the researcher.

Information Theory   A statistical and mathematical theory of communication dealing with the nature, effectiveness, and accuracy of transmitting information between humans, between machines, and between humans and machines.

Informed Consent   Principles for protecting human research subjects based on the Nuremberg Code of the 1940s and the Helsinki Declaration of the 1990s. In brief, potential subjects must be informed, prior to any treatment, of the possible risks of participation and that they may withdraw at any time.

Institute for Social Research (ISR)   Founded by Rensis Likert and located at the University of Michigan, the ISR is perhaps best known as the parent organization of the Survey Research Center (SRC).

Institutional Research   What researchers in higher education often call *evaluation research, especially research on their own institutions. The research is usually designed to help administrators of a higher education institution plan and make decisions.

Institutional Review Board (IRB)   A screening panel for research projects that must, by law, be in place at any U.S. institution receiving federal funding. Its focus is protecting human subjects from potential harm caused by research. See *research ethics.

Instrument   Any means used to measure or otherwise study subjects. In the language of social and behavioral research, an instrument can call to mind a mechanical device (as it does in ordinary language—a dentist's drill, a saxophone), but it is used more broadly to include written instruments, such as attitude *scales or *interview schedules.

Instrumentalism   The philosophical belief that *theories are not statements of truth about the real world, but are rather mere tools or *instruments that allow scientists to make predictions. This *positivist belief has been sharply criticized by *qualitative researchers.

Instrumental Variable   A variable used to replace an independent variable in a *regression equation when the independent variable is highly correlated with the *error term. A good instrumental variable will be highly correlated with the independent variable it is replacing but will not be correlated with the error term.

Integer   A whole number, whether negative or positive: 1, 2, 4, and 5 are integers; 1.2 and 4/5 are not.

Interaction Effect   The joint effect of two or more *independent variables on a *dependent variable. Interaction effects occur when *independent variables not only have separate effects, but also have combined effects that are different from the simple sum of their separate effects. In other terms, interaction effects occur when the relation between two variables differs depending on the value of another variable. The presence of statistically significant interaction effects makes it difficult to interpret *main effects. Also called "conditioning effect," "contingency effect," "joint effect," and "moderating effect." Compare *additive relation, *moderator variable.

When two variables interact, this is called a *first-order interaction; when three interact, it is a second-order interaction, and so on. Interaction effects may be *ordinal or *disordinal; see the entries under those terms for more details.

For example, suppose a cholesterol reduction clinic had two diets and one exercise plan. Exercise alone was effective and dieting alone was effective. As can be seen in the following graph, for patients who did not exercise, the two diets worked about equally well. Those who went on Diet A and exercised got the benefits of both. But those who combined exercise with Diet B got a bonus, an interaction effect. All patients could benefit by dieting and exercising, but those who followed Diet B and exercised benefited more.

Another instance of an interaction effect is when the treatments or independent variables interact with *attributes of the people being studied, such as their age or sex; these interactions are often referred to as *moderating effects. The diet and exercise example illustrates the interaction of two treatments.

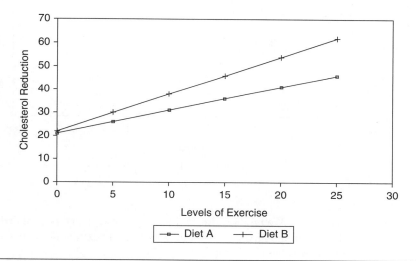

Figure I.1   Interaction Effect

Intercept   The point at which a *regression line crosses (or "intercepts") the vertical (y) axis, that is, when the value on the x (horizontal) axis is zero. Put another way, the intercept is the expected value of the *dependent variable when the value of the *independent variable is zero. Also called "y intercept." See *regression constant and *constant, definition (c).

Regression lines often cross the x axis, thus producing an x intercept, but this is rarely discussed. In *multiple regression, the y intercept is the mean value of the *dependent variable for a case with a value of zero on all the *independent variables.

If we had 100 students' scores on an exam and surveyed them regarding how many hours they spent studying for the exam, we could make a scatter plot of the data and draw the regression line as in the following. The point at which the line crosses the vertical axis is the intercept. It is around 70. The regression coefficient or *slope is .8, which means that on average for every hour a student studied, his or her estimated score went up, beyond the intercept, a little less than a point. Note that no student actually claimed not

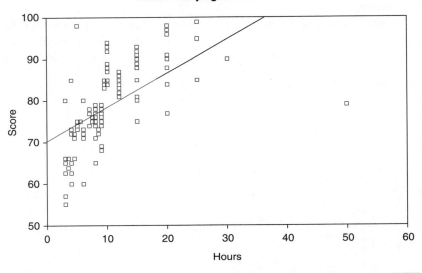

Figure I.2   Intercept

to have studied at all. We might estimate that a student who had not studied at all would score 70 on the exam, but that would be an *extrapolation from the data, not a real value in the data set.

Interclass Correlation   An ordinary *correlation. The term *inter*class is used to distinguish it from an **intra*class correlation (ICC).

Interclass Variance   Another term for *between-groups differences in *ANOVA. By extension, *intra*class variance refers to within-groups differences.

Intercorrelation   A correlation between *independent variables—as contrasted with a correlation between an independent variable and a dependent variable. See *multicollinearity.

Intermediary Variable   Another term for *mediating variable or *intervening variable.

Internal Consistency   The extent to which items in a *scale are *correlated with one another, and by extension, the extent to which they measure the same thing. See *Cronbach's alpha.

Internal Reliability   The extent to which items in an *index or *scale (such as questions on a test of a particular *domain of knowledge) give similar results. See *Cronbach's alpha, *reliability coefficient.

Internal Validity   The extent to which the results of a study (usually an *experiment) can be attributed to the *treatments rather than to flaws in the research design. In other words, internal validity is the degree to which one can draw valid conclusions about the causal effects of one variable on another. It depends on the extent to which *extraneous variables have been controlled by the researcher. Compare *external validity.

Interpolation   The act of estimating an unknown value by using its position among a series of known values. Compare *extrapolation.

   For example, if the average weight of 5-year-olds in a sample were 50 pounds, and the average weight of 7-year-olds were 70 pounds, we might interpolate that 60 pounds would be the average weight of 6-year-olds in the sample.

Interquartile Range (IQR)   A measure of *dispersion calculated by taking the difference between the first and third *quartiles (that is, the 25th and 75th percentiles). In short, the IQR is the middle half of a distribution. Also called "midspread." See *box plot, *hinge. Compare *standard deviation.

Interrater Reliability   Agreement or consistency among raters; the extent to which raters judge phenomena in the same way. Ratings often involve assigning numbers to qualitative assessments. Olympic judges provide a familiar example. Although all the judges of an Olympic performance rarely award it exactly the same score, a very high level of agreement (interrater reliability) is common. Interrater reliabilities are usually measured by *correlation coefficients. See *Cohen's kappa.

Interrupted Time-Series Design   An approach that requires researchers to examine trends in the *dependent variable before, during, and after an intervention or *treatment. The purpose is to avoid such *threats to validity as *history effects and *pretest sensitizing. The method is often used in clinical settings to study a single subject.

   Suppose, for example, that the management of a company thinks that productivity will go up if employees attend a special training seminar (the *treatment). To see whether the seminar is effective, the following design is used. Repeated pretesting (Weeks 1–3) establishes a *baseline and helps reduce *Hawthorne effect bias. Repeated posttesting helps establish whether any improvements in productivity last beyond the first posttest week (Week 5).

Table I.1   Interrupted Time-Series Design

Week 1	Week 2	Week 3	Week 4	Week 5	Week 6	Week 7
Pretest	Pretest	Pretest	Treatment	Posttest	Posttest	Posttest

Intersection   In *set theory, the overlapping of two or more sets; said of the elements shared by subsets. Symbolized "*cap," as in A ∩ B. See *Venn diagram.

Intersubjective Agreement   A state that exists when people (subjects) agree. It is a main criterion of *objectivity.

Interval Estimate   An estimate that includes a range of scores. Compare *point estimate. See *confidence interval.

For example, the following statement contains an interval estimate: "If the election were held today, we estimate that Candidate A would get between 51% and 57% of the vote." By contrast, a point estimate would read: "If the election were held today Candidate A would get 54% of the vote."

Interval Scale (or Level of Measurement)   A scale or measurement that describes variables in such a way that the distance between any two adjacent units of measurement (or "intervals") is the same, but in which there is no meaningful zero point. Scores on an interval scale can meaningfully be added and subtracted, but not multiplied and divided. Compare *ratio scale.

For example, the Fahrenheit temperature scale is an interval scale because the difference, or interval, between (say) 72 and 73 degrees is the same as that between 20 below 0 and 21 below 0. Since there is no true zero point (zero is just a line on the thermometer), it is an interval, not a ratio scale. There is a zero on the thermometer, of course, but it is not a true zero; when it's zero degrees outside, there is still some warmth, more than when it's 20 below.

To take another example, if on a 20-item vocabulary test Mr. A got 12 correct and Mr. B got 6 right, it would be correct to say that A answered two times as many correctly, but it would not be correct to say that A's vocabulary was twice as large as B's—unless the test measured all vocabulary knowledge and getting a zero on it meant that a person had no vocabulary at all (in that case the test would be an example of a ratio scale).

Intervening Variable   A variable that explains a relation, or provides a causal link, between other variables. Also called "mediating variable" and "intermediary variable." Compare *moderating variable, *suppressor variable.

For example, the statistical association between income and longevity needs to be explained, since having money by itself does not make one live longer. Other variables intervene between money and long life. For instance, people with high incomes tend to have better medical care than those with low incomes. Medical care is an intervening variable; it mediates the relation between income and longevity. Figure I.3 represents this graphically.

Income → Medical Care → Longevity

Figure I.3   Intervening Variable

Interviewer Effects   Influences on the responses of persons being interviewed that stem from the characteristics of interviewers. Respondents may answer questions differently depending upon the accent, age, gender, and so on of the person asking the questions. Few systematic studies of such effects have been conducted, but most of these have found the effects to be quite modest.

Interview Protocol   A list of questions and instructions for how to ask the questions. It is used as a guide when interviewing subjects.

Interview Schedule   A list of questions and spaces to write down the answers. It is used by interviewers to record respondents' answers. Compare *questionnaire.

Intraclass Correlation (ICC)   A measure of homogeneity among the members of a group, class, or cluster. It is often used as a measure of *inter-rater reliability. Several different procedures for calculating ICCs exist; they are used for measuring reliability in different circumstances, such as different numbers of raters. The ICC is used in *hierarchical linear models to indicate the proportion of the total variance attributable to the higher level variables. See *correlation.

Intraclass Variance   Another term for *within-group differences in *analysis of variance. See *interclass variance.

Invariance   The condition of being unchanged by specific mathematical *transformations, as an invariant *factor. For example, if you double the values of two variables, the *correlation between them does not change; it is invariant.

Inverse Relation (or Correlation)   A relation between two variables such that when one goes up the other goes down, and vice versa. Also called *negative relation. Compare *direct relationship and *positive relation.

   For example, the relation between the female employment and fertility rates tends to be inverse (or negative): The higher the female employment rate the lower the fertility rate. See *linear relation for another example.

Ipsative Measure (or Scale)   A *rank-order scale in which a particular rank can be used only once. The opposite is usually called a "normative" scale.

   For example, if you gave raters the following instructions, the results would be an ipsative scale: "Here are 11 movies; rank them from best to worst, giving the best a 10, the next best a 9, the next an 8, and so on down to the worst, which gets a 0." By contrast, if you said, "Here are 11 movies; rate them on a scale of 0 to 10," you would be asking for a normative scale. One rater could think the movies were all excellent and give them all 10s and 9s; another might believe they were terrible and give them all 1s and 2s.

   When several raters uses an ipsative scale, the means, medians, and standard deviations of their rankings are always the same. But, when several raters use a normative scale, this is not necessarily (and rarely is) the case.

IQR   *Interquartile range.

IRB   *Institutional Review Board.

IRT   *Item response theory.

Irrational Number   A number with infinite, nonrepeating decimals such as pi or the square root of 2 (1.414213562 . . .). It is said to be irrational, not because it is wacky, but because it cannot be exactly expressed as a *ratio of whole numbers. For example, 22/7 comes close to, but does not exactly equal pi (3.14159265 . . .). Compare *rational number.

Isomorphic   Having a form or structure similar to something else. Said of a *theory that can be deduced from another theory because the two are logically equivalent. Said of measurements when big units of measurement are used to measure big things and small units are used to measure small things.

   For example, we don't measure the distance from Los Angeles to San Francisco in inches; we could, but we don't, because the measurement scale and the thing measured would not be isomorphic.

Item Analysis   In testing research, the study of questions (test items) based on individuals' responses to the questions. See *item response theory, *Rasch modeling.

Item Response Theory (IRT)   A group of methods designed to assess the *reliability, *validity, and difficulty of items on tests. The assumption is that each of the items is measuring some aspect of the same underlying (*latent) ability, trait, or attitude. IRT is important for determining equivalency of tests (such as two versions of the same proficiency exam) as well as for determining individuals' scores. IRT uses a version of *logistic regression as its basic tool, with the dependent variable being the log of the odds of answering questions correctly. See *differential item functioning and *Rasch modeling, which is an alternative approach to the same class of research problems.

Iteration   Generally, a repetition. (a) Any procedure in computation in which a set of operations is repeated, such as each time an *estimate is obtained in a series of estimates. (b) A method of successive approximation in which each step is based on the results of the preceding steps.

IV   Abbreviation for *independent variable.

Jackknife Method   A way to estimate *standard errors and *confidence intervals. The basic approach is to take repeated random subsamples, without replacement, of one's original sample, eliminating one or more data observations each time. The main advantage of the method is that it requires no assumptions about *underlying distributions; it is thus a *nonparametric or *distribution-free method. See *bootstrap methods, which are an extension of the jackknife, and *resampling.

J Curve (or J-Shaped Distribution)   (a) A curve describing the relation between a nation's balance of payments (imports relative to exports) and the value of its currency; as the value of the currency goes down, the balance of payments goes down for a while but then goes up, slowly at first, then rapidly, which results in the J-shape.

   (b) A curve describing a *frequency distribution that looks roughly like an uppercase letter J, that is, with minimum frequency at low levels of the *x axis and rapidly increasing frequencies at higher levels. Examples of such curves are often found illustrating the frequency of adherence to a norm or compliance with a standard of behavior. (Depending on how a relationship is graphed, the J may be in its normal upright position or, as is often the case, lying on its side.)

Jittering   A method of making *scatter diagrams easier to interpret when many data points are identical. It involves adding a tiny amount of random variation to the points so that they are not on top of one another and can be seen.

John Henry Effect   A tendency of persons in a *control group (those who are not receiving an experimental *treatment) to take the experimental situation as a challenge and exert more effort than they otherwise would; they try to beat those in the *experimental group. This, of course, negates the whole purpose of having a control group.

For example, in order to see if a new power tool were worth the investment, a supervisor in a construction firm might provide some workers (the experimental group) with the new power tool; the rest of the workers (control group) continue using the old tool. The workers using the old tool might work much harder to show that they were just as good and should get the new tool, too. They might actually produce more, even though, under ordinary conditions (not influenced by the John Henry effect) workers using the new tool would be more productive.

Joint Contingency Table   A table illustrating how two or more *independent variables jointly affect a *dependent variable.

For example, the following joint contingency table shows the unemployment rate (dependent variable) of different age and race groups over time (independent variables). Unemployment is jointly affected by age and race and year. The table shows that African Americans and young persons were more likely to be unemployed and that rates for all groups went down in the 1990s and back up after 2000.

Table J.1   Joint Contingency Table: Unemployment Rates (1994–2004)

	Whites		Blacks	
Year	All	Aged 16–19	All	Aged 16–19
1994	5.7	16.0	13.1	32.7
1996	4.9	15.4	10.6	33.5
1998	4.0	11.8	9.4	29.7
2000	3.4	11.1	8.0	23.1
2002	5.1	14.5	10.0	31.3
2004	4.9	14.1	10.5	32.9

NOTE: Figures are for January of each year.

Jointly Independent   Two variables (X and Y) are said to be jointly independent of a third variable (Z) if their *joint probability distribution is uninfluenced by the value of Z.

Joint Probability   The probability of two or more *conditional events occurring together.

For example, the probability of drawing a card of clubs from a normal deck is 1 out of 4 (or .25). The chances of drawing a 7 are 1 out of 13 (or .0769). The probability of drawing a card that is both a club and a 7 is 1 out of 52 ($1/4 \times 1/13 = 1/52$) or .01923 ($.25 \times .0769$).

Judgment Sampling   A procedure in which a researcher makes a judgment that a *convenience sample (e.g., volunteers) might be similar enough to a *random sample that it could make sense to use statistical procedures designed for use on random samples. Selecting a sample according to the researcher's judgment of its representativeness is recommended only when a *probability sample is impossible or highly impractical, as it often is. See *purposive sample.

Just-Identified Model   Another term for a *fully recursive, but not *over-identified, model. A model in which the number of variables and the number of parameters to be estimated are equal. Also called a *saturated model. See *recursive model.

J

K (a) Uppercase *K,* the usual symbol for a *coefficient of alienation. (b) Lowercase *k,* common symbol for the number of groups or *samples in a study. (c) Symbol for *kernel.

Kappa [K, κ] See *Cohen's kappa.

Kappa Coefficient See *Cohen's kappa, which is a measure of *inter-rater reliability.

Kendall's Coefficient of Concordance A *nonparametric statistical test of the agreement among sets of rankings. Symbolized *W. W* can range from 0 (no agreement) to 1.0 (complete agreement). See *interrater reliability.

For example, if we wanted to see how much seven wine tasters agreed (were in "concord") about their rankings of a dozen different wines, we could use Kendall's coefficient.

Kendall's (Tau) Correlations One of three measures of *association (tau a, tau b, and tau c) between two *ordinal variables. This *correlation between ordinal variables is used when the ranks of the ordered categories are not treated as interval scales. It is generally considered better than *Spearman's rho because of the way it deals with tied ranks. Compare *Somers's *d.*

Kernel Methods Methods for estimating *probability density functions. The kernel occurs in the integral equations of calculus.

Kilobyte A measure of computer memory. Literally, one thousand *bytes, but actually 1,024 bytes.

Kim's *d* A measure of *association between two *ordinal variables used when the ranks of the ordered categories are not treated as *interval scales. Contrast *Spearman's rho.

K

KMO Test   Kaiser-Myer-Olkin test. An indicator of the strength of relationships among variables in a correlation matrix. It is determined by calculating the correlations between each pair of variables after *controlling for the effects of all other variables. Compare *Bartlett's test. The KMO statistic can range from 0 to 1.0; .70 is often considered a minimum for conducting a *factor analysis.

Kolmogorov-Smirnov Tests   *Nonparametric tests (for *ordinal data) of whether two distributions differ and whether two samples may reasonably be assumed to come from the same population; they are *goodness-of-fit tests.

KR20 and KR21   Abbreviations for Kuder-Richardson formulas 20 and 21.

Kruskal-Wallis Test   A *nonparametric test of *statistical significance used when testing more than two independent samples. It is an extension of the *Mann-Whitney U test, and of the *Wilcoxon test, to three or more independent samples. It is a nonparametric, *one-way ANOVA for *rank-order data and is based on *medians rather than *means. Symbolized H.

Kuder-Richardson Formulas (20 and 21)   Measures of the internal consistency or *reliability of tests in which items have only two possible answers—such as agree/disagree or yes/no. Formula 21 is a simplified and easily computable version of Formula 20. Both give conservative (low) estimates of a test's reliability. KR20 estimates the outcome of taking all possible orders (*permutations) of questions on a test, computing a *split-half reliability coefficient for each of these orders, and finally taking the means of those coefficients. Compare *Spearman-Brown formula and *Cronbach's alpha.

Kurtosis   An indication of the extent to which a distribution departs from the bell-shaped or *normal curve by being either pointier (leptokurtosis) or flatter (platykurtosis).

   Kurtosis can be expressed numerically as well as graphically. Computer programs often provide such numbers. The basic rule for interpreting them is that negative numbers mean flatter than normal and positive numbers mean more peaked than normal. The number for a normal distribution is zero.

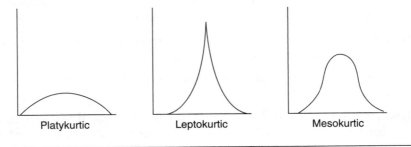

Figure K.1   Kurtosis

L   Symbol for *logit.

L²   Symbol for the *maximum likelihood ratio chi-squared statistic.

Laboratory Research   Any of several methods of isolating subjects so as to *control *extraneous variables. Often considered to be synonymous with *experimental research. The advantage of the laboratory method in the social and behavioral sciences is also its disadvantage. Subjects can be isolated from contexts that might influence their behavior; hence the researcher can focus on those *independent variables of interest. However, it may be difficult to generalize results to situations outside of the laboratory, since people seldom act in isolation from context. In other words, using laboratory research sometimes means trading a gain in *internal validity for a loss in *external validity. See *randomized clinical trial.

Lagged Dependent Variable   Said of a dependent variable whose value at a particular time is to some degree dependent on its value at a previous time or dependent on the values of independent variables at an earlier time.

For example, the amount of money families spend annually on vacations (the dependent variable) may tend to fluctuate with their income (an *independent variable). If income goes up, the amount spent on vacations may go up. But the amount families spend this year also tends to be predictable by how much they spent in previous years. For instance, families that didn't spend much on vacations in the past might continue, for some time, to use any increases in income for other things. This means that one of the predictors of the value of this year's dependent variable (spending on vacations), is last year's spending. In other words, last year's dependent variable becomes one of this year's independent variables.

L

Lagging Indicator   An indicator in which general trends show up only after a time, lagging behind other indicators. Contrast *leading indicators.

For example, one common lagging indicator is the average length of time individuals are unemployed. It goes down as the economy gets stronger, but not right away.

Lambda [Λ, λ]   (a) The usual symbol for *coefficients transformed by taking their natural *logarithms. See *log-linear analysis, *transformation. Compare *Wilks's lambda. (b) Common symbol for *eigenvalue. (c) Goodman and Krusdal's lambda, which is a *measure of association appropriate to use when the *variables being described are *categorical (*nominal or *discrete). Lambdas range from zero, when knowing one variable tells you nothing about another, to 1.0, when knowing one always enables you to predict the other. Lambda is a *PRE measure.

For an example of definition (c), say that a statistics professor had 40 students in a class, 20 women and 20 men. The professor gave a pass/fail test; 22 students passed and 18 failed. If, of the 22 students who passed, 11 were men and 11 were women, there would be no association between the two variables; knowing the sex of students wouldn't help predict whether they passed or not. The lambda would be zero. But suppose that 20 of the 22 who passed were female and 2 were male (and therefore all 18 who failed were male). Then there would be a strong association between sex and success on the test. Knowing students' sex would most often enable us to tell whether they would pass or fail. The lambda in this example is .89. This means that our prediction is 89% better when we know the students' sex.

Laspeyres Index   One of the earliest methods of calculating *index numbers and *base years for *time-series indexes, particularly for tracking prices and measuring inflation. The formula is still used today to compute the *consumer price index (CPI) in the United States. A standard "basket" of goods is purchased and its cost is compared with the cost of the same basket at a baseline date. The index is a *weighted average, with the prices of some goods more heavily determining the index number.

Last In First Out (LIFO)   Compare *first in first out, *first in last out.

Latent Class Analysis (LCA)   A method similar to *factor analysis, but used with *categorical data. While factor analysis is used to discover *latent variables, LCA is used to find latent categories or "classes" of variables.

Latent Class Models   Models designed to discover whether complex relations between observed categorical variables can be explained by relationships between unobserved or *latent variables.

Latent Curve Models   See *growth curve analysis.

Latent Factor   See *latent variable.

Latent Function   In *functionalism, a purpose or use of a social phenomenon that is not obvious (it is hidden or "latent") to social actors. Researchers
hypothesize its existence to explain otherwise mysterious phenomena.
Researchers looking for latent functions are looking for *latent variables.

    Men's neckties might be a good example. At one time they were pre-
sumably scarves meant to keep one warm. But today they are worn indoors
in well-heated buildings, and even on very hot days when it is uncomfort-
able to do so.

    Keeping warm cannot be the function they fill; it cannot explain their
widespread use. Nor can neckties be wholly explained by their decorative
functions, since there are many ways men could decorate themselves (such
as wearing a brooch pinned to the collar). But all except neckties are considered socially inappropriate—for men, but not women, at least in some
circumstances. So what is the latent function of necktie wearing? Latent
functions are always speculative because, like all latent variables, they are
hard to study and measure directly. But one might hypothesize that by
wearing a necktie, a man makes the following kind of statement: "I am a
serious person; I recognize that this is an important social context (work, a
formal social event); and by dressing appropriately, I show you that I am
the kind of person who can be trusted to do the right thing. Were I not wearing a tie you might imagine that I am frivolous or rebellious."

Latent Structure   A pattern of relations among variables that is not directly
observable but is hypothesized to exist so as to explain variables that are
observable. See *latent variable, *factor analysis, *structural equation models.

Latent Variable   An underlying characteristic that cannot be observed or
measured directly; it is hypothesized to exist so as to explain variables,
such as behavior, that can be observed (*manifest variables). Latent variables are also often called *factors, especially in the context of *factor
analysis. Compare *latent function, *latent class analysis, *structural equation models.

    For example, if we observed the votes of members of a legislature on
spending bills for the military, medical care, nutrition programs, education,
law enforcement, and promoting business investment, we might find underlying patterns that could be explained by postulating latent variables
(factors), such as conservatism and liberalism.

Latin Square   A method of allocating subjects, in a *within-subjects experiment, to *treatment group orders. So called because the treatments are
symbolized by Latin (not Greek) letters. The main goal of using Latin
squares is to avoid *order effects by rotating the order of treatments. See
*counterbalancing, *Greco-Latin square.

In the following example, A, B, C, and D are treatments. There are four subjects and four orders of treatment. Subject 1 would receive the treatments ABCD in that order, the order for Subject 2 would be BDCA, and so on. Note that a Latin square must be square, i.e., the number of rows and columns must be equal. The number of subjects must equal or be a multiple of the number of treatments—in this example, 4, 8, 12, 16, and so forth.

Table L.1   Latin Square

	Order			
	1st	*2nd*	*3rd*	*4th*
Subject 1	A	B	C	D
Subject 2	B	D	A	C
Subject 3	C	A	D	B
Subject 4	D	C	B	A

Law   A statement about the relations among *variables that has been frequently confirmed and that seems to hold under all circumstances. While a law is generally thought to be more certain than a theory, the difference between "law" and "theory" is often little more than a matter of accidents of usage (for example, the law of supply and demand, but the theory of evolution). Usage varies considerably. Sometimes law is used to mean universal *empirical generalization; in that usage, a law would describe but not explain the regularity; a theory would explain it.

   Modern social and behavioral scientists rarely refer to their generalizations as laws; they use "theory" almost to the exclusion of "law" to refer to statements about regular relations among variables.

Law of Averages   The principle that *random errors in measurement will tend to balance one another out, that is, they will as often be above as below the true values. Also, as *sample size increases, the sample *statistic will be a better estimate of the population *parameter. Compare *law of large numbers, *central limit theorem.

Law of Large Numbers   Description of how larger *samples are better (more representative) of the *populations from which they were drawn—specifically, the larger the sample, the more likely it is that the sample *mean will equal the population mean. This law holds only if the larger samples are not *biased; increasing the size of a biased sample leads to little or no improvement. Compare *central limit theorem.

Leading Indicators   Events likely to precede other events that researchers want to *forecast. Usually used in the phrase "leading *economic* indicators," but the concept is also useful in other fields where *time-series data are used. Contrast *lagging indicator.

Learning Curve   The tendency to learn how to do something more efficiently the more often you do it. When graphed, this yields different kinds of curves—steep and not steep, for tasks that are and are not likely to be learned quickly with repetition. The concept is widely used in manufacturing, where the focus is on the unit cost of production, but it can be applied to other sorts of learning as well.

Least Squares Criterion (or Principle)   The rule that using the *mean to predict the scores in a distribution results in predictions that are most accurate, with "accurate" in this case indicating that using the mean yields the smallest possible sum of squared *errors (or squared *deviation scores). In *regression analysis, it is called *ordinary least squares (OLS), which is a method or criterion for calculating the *regression equation (or drawing the regression line) that best summarizes or fits a distribution. See *general linear model, *generalized least squares.

Lemma   A *theorem that has been proven and then used to prove another theorem.

Leptokurtic   More peaked than a *normal curve. See *kurtosis for an illustration.

Level   A *treatment or a *condition of an *independent variable (IV) in an experiment. A particular value of an IV in a study. "Level" implies amount or magnitude in ordinary language, and it is used that way in experiments, too. If subjects were given 10 cc, 15 cc, or 20 cc of a medication, those would be the three levels. But "level" is also used for *categorical variables, such as medications A, B, and C, where the three are different in kind, not different in amounts of the same thing.

Level of Analysis (or Aggregation)   If we were to study the United States we could look at individuals, neighborhoods, counties, states, or regions. Individuals would be the lowest level of analysis or aggregation; regions would be the highest. The lower the level of analysis, the higher the level of specificity tends to be, and vice versa. See *level of generality.

Level of Generality   The breadth of generalizations.

Take, for example, statements that apply to deviance, crime, and theft. Theft, a specific type of crime, is at the lowest level. Crime is a more general category than theft, but is less broad than deviance, which includes crimes, but can also be taken to mean any departure from the ordinary.

Table L.2   Level of Generality

Level	Example
Low	Theft
Middle	Crime
High	Deviance

Level of Measurement   A term used to describe measurement scales in terms of how much information they convey about the differences among values—the higher the level, the more information.

There are four levels of measurement. Arranged in order of strength, from the highest to the lowest, they are: *ratio, *interval, *ordinal, and *nominal. It is possible to describe data gathered at a higher level with a lower level of measurement; but the reverse is not possible. For example, one can express income in dollars and cents (interval level) or with ordinal descriptions like upper, middle, and lower class.

It is important to be aware of the level of measurement you are using, because statistical techniques appropriate at one level might produce ridiculous results at another. For example, in a study of religious affiliation, you might number your variables as follows: 1 = Catholic, 2 = Jewish, 3 = Protestant, 4 = Other, 5 = None. The religion variable is measured at the nominal level. The numbers are just convenient labels or names; one cannot treat them as if they mean something at the interval level; one cannot add together a Jewish person (2) and a Protestant person (3) to get an atheist (5).

Considerable controversy exists concerning which statistics can validly be used to analyze variables measured at different levels of measurement. The debates usually revolve around questions of how serious a distortion occurs when one violates particular *assumptions presumed by certain statistical techniques. As with constitutional law, there are in statistics strict and loose constructionists in the interpretation of adherence to assumptions.

Level of Significance   More fully, the level of *statistical significance. The probability that a result would be produced by chance (*sampling error, *random error) alone.

The level of significance indicates the risk or *probability of committing an error (*Type I error in *hypothesis testing). The level of significance is stated as a probability, often abbreviated p, followed by a number, such as, $p < .05$ or $p < .01$. The smaller the number, the smaller the chance of Type I error, and the more statistically significant the finding. See also *alpha level, *p value.

Note: The level of statistical significance says nothing about a finding's *substantive or *practical significance; there are no statistical tests for substantive or practical significance.

Level of Specificity   See *level of analysis, *level of generality.

Levene's Test   A test for *homogeneity (equality) of variances in distributions. Often used prior to conducting an *ANOVA and in interpreting the results of a *t-test. The importance of Levene's test lies in the fact that the ANOVA and the t-test (among other statistical techniques) assume equality of variances. An ANOVA or a t-test is not valid if the assumption is not met.

Leverage Point   In *regression diagnostics, an *outlier among the *independent (*predictor) variables that is large enough to significantly affect the interpretation of the regression equation.

Liar's Paradox   A paradox that arises from someone making a statement such as: "I am lying," or "This statement is false." If the statement is true, then it is false; and if it is false, then it is true.

Life Expectancy   The predicted number of years yet to be lived by persons of a particular age. The older the person, the shorter the life expectancy. This is calculated by taking the average number of years lived by persons in a particular birth *cohort at a given time or time interval. Life expectancies are based on *cross-sectional data, which means that using them to make predictions about future life expectancies is based on the assumption that future rates will be affected by the same variables as those at work at the time the cross-section was taken. See *life table.

Life Table   A table showing *life expectancy at various dates and for different groups, usually birth *cohorts.

For example, the following simplified life table follows the life, death, and life expectancies per 100,000 for a hypothetical (but not unrealisitc) population. In row 1 we start with 100,000 persons, aged zero to 1 year, of whom 650 die by the end of the period. The survivors are expected to live another 79 years. 100,000 minus 650 leaves 99,350 survivors in row 2, of whom 150 die. Those who make it to the end of the period are expected to live 78 years. Many things can be seen in this table. For example, infancy (years 00 to 01) is much more hazardous than early childhood (ages 01 to 05). If children make it through their first year they are quite unlikely to die until adulthood. It is not until people are much older that the death rate again approaches that of infancy. Looking at a later point in the age spectrum, in row 16, we see that people who survive that period can expect to live another 15 years on average.

Table L.3   Life Table

A. Number of Years (Age)	B. Number Living	C. Number Dying	D. Expected Years to Live
1. 00–01	100,000	650	79
2. 01–05	99,350	150	78
3. 05–10	99,200	80	75
4. 10–15	99,120	90	70
5. 15–20	99,030	220	65
6. 20–25	98,810	240	60
7. 25–30	98,570	300	55
8. 30–35	98,270	430	50
9. 35–40	97,840	600	46
10. 40–45	97,240	840	41
11. 45–50	96,400	1,200	36
12. 50–55	95,200	2,000	32
13. 55–60	93,200	3,000	27
14. 60–65	90,200	5,000	23
15. 65–70	85,200	7,000	19
16. 70–75	78,200	9,600	15

LIFO   *Last in-first out.

Likelihood (L)   The *probability of observed results (in a *sample) given the estimates of the *population parameters. In other words, the *conditional probability of observed frequencies or values given expected frequencies or values. It is perhaps most widely used as a way to determine the *goodness-of-fit of a *logistic regression model. See *likelihood ratio, *maximum likelihood estimation.

Likelihood Ratio (LR)   As the name implies, the LR is a *ratio of two *likelihoods, specifically the ratio of two likelihoods given two different hypotheses about the data. It is the ratio of the outcome if the *alternative hypothesis is true to the outcome if the *null hypothesis is true. It is widely used as a *test statistic for comparing models, specifically two *nested models. The smaller the LR, the stronger the relationship.

Likelihood Ratio Chi-Square   Short for *maximum likelihood chi-square.

Likert Scale   A widely used questionnaire format developed by R. Likert. Respondents are given statements and asked to respond by saying whether they "strongly agree," "agree," "disagree," or "strongly disagree." Wording varies considerably; for example, people might be asked if they "totally approve," "approve somewhat," and so on. See *summated scale.

Likert scales, and Likert-like scales, are the most widely used attitude scale types in the social sciences. They are comparatively easy to construct, can deal with attitudes of more than one dimension, and tend to have high reliabilities.

Lilliefors Test   A *test statistic for the hypothesis that a sample has been drawn from a population that is normally distributed. It is a modification of the *Kolmogorov-Smirnov test.

Limit   (a) In mathematics, a theoretical end point that can be ever more closely approached, but never quite reached. For example, if we added the fractions $1/2 + 1/4 + 1/8 + 1/16 + 1/32$ . . . and so on, each time adding half of the previous fraction to the string, the more we added the closer we would get to the limit of 1.0; but we would never reach it. Compare *asymptote.

   (b) In *probability theory, the larger the number of *trials, the closer the *empirical probability gets to the limit or the *theoretical probability. The more times we flipped a fair coin, the closer the proportion of heads would get to the limit of .5, or the closer the ratio of heads to tails would get to 1:1.

Lindquist Type I ANOVA   A *two-way, *mixed-design ANOVA, that is, an ANOVA with one repeated measure, a one-time measure.

Linear   Of, relating to, or resembling a straight line.

Linear Algebra   A broad category of algebra including *matrix algebra.

Linear Combination   If when two variables are graphed they form a straight line, they are called a linear combination. A *dependent variable is a linear combination of two or more independent variables when a *linear equation can be used to describe the relation between the dependent variable and the independent variables.

Linear Dependence   Said of a variable that depends on another, in the sense that it can be derived from the other using a *linear equation. When *independent variables have a great deal of linear dependency, this raises the problem of *collinearity. *See *function.

Linear Equation   An equation that can be plotted on a graph as a straight line. Such an equation contains no *powers higher than 1, and the variables (a and b in the following example) are combined by addition or subtraction, not by multiplication or division.

   For example, $y = a + 2b$ is linear, but $y = a + b^2$ is not, nor is $y = a \times 2b$.

Linear Function   A *linear relation expressed as a *linear equation, that is, one that does not contain terms with powers or interaction terms.

Linear Regression Analysis   A method of describing the relationship between two or more variables by calculating a "best-fitting" straight line (or plane) on a graph. The line averages or summarizes the relationship. The result is a *regression line, which can also be expressed in a *regression equation.

L

*Regression analysis is a generic term; without further specification, it almost always means the linear variety. Compare *curvilinear relation, *polynomial regression, *logistic regression.

Linear Relation (or Correlation)   A relationship that, when plotted on a graph, forms a straight line. It forms a straight line because the direction and the rate of change in one variable are constant with respect to changes in the other. Compare *curvilinear relation, *monotonic relation.

For example, if a baker notices that whenever he raises the price of a loaf of bread by a nickel, sales drop by 10 loaves, and every time he lowers the price by a nickel, sales go up 10 loaves, the relationship between price and sales (graphed below) is linear. This is also an example of an *inverse (or negative) relation, since when one variable goes up the other goes down. Such relations are seldom perfectly linear, as in this example. The *linear equations used to describe real-world relations are approximations. (Note that the baker makes a bigger profit selling 130 loaves for $1.00 than 250 loaves at 40 cents each.)

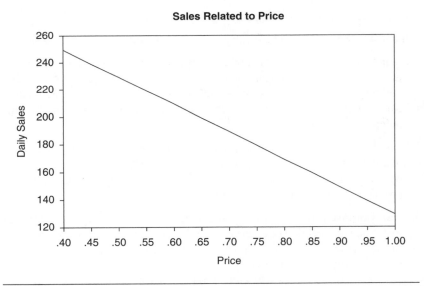

Sales Related to Price

Figure L.1   Linear Relation (or Correlation)

Linear Transformation   Changing a number, group of numbers, or an equation by adding, subtracting, multiplying, or dividing by a constant. The best known example is probably multiplying *proportions by 100 (the constant) in order to change ("transform") them into percentages. Called "linear" because when you plot the old values against the new on a graph, the result is a straight line. Compare *nonlinear transformation and see that entry for an illustration.

Link Function   In *generalized linear models, the *transformation of the *dependent variable. It connects (links) the nonlinear *outcome variable with the *linear combination of the *predictor variables.

LISREL   Linear Structural Relations. A *computer program used for analyzing *structural equation models (SEM). The software brand name has become so well known that it is sometimes used for the methods of analysis as well as for the technology for executing them. Compare *EQS.

LISREL can be used to analyze causal models with multiple indicators of *latent variables and structural relations among latent variables. This analysis (SEM) is more powerful than simple *path analysis, because bias due to random measurement error in the multiple indicators of the latent variables is corrected. It goes beyond the more typical *exploratory factor analysis, and allows the researcher to do *confirmatory factor analysis. See *canonical correlation analysis.

List Sample   Another term for *systematic sample.

Listwise Deletion   Completely removing a case from the calculation of *coefficients when that case has any data missing. Only cases with values for all variables are used. Also called "casewise deletion." Compare *pairwise deletion, *mean substitution.

For example, consider the following *data matrix with four cases and three variables in which there is no data for Case 1 on Variable 1. Computer programs using listwise deletion will not use case 1 at all, whereas those using pairwise deletion will use it to compute a correlation between Variable 2 and Variable 3. Pairwise deletion is often preferred because it retains more data for use in calculations. A third option is *mean substitution. In this example, the mean score of the other cases on Variable 1 (11.67) would be used as an estimate of the missing number. Often a researcher will report the results of all three procedures.

Table L.4   Listwise Deletion

	Var. 1	Var. 2	Var. 3
Case 1	—	15	40
Case 2	10	6	42
Case 3	12	17	48
Case 4	13	20	47

Literature Review   A survey and interpretation of the research findings (the "literature") on a particular topic, usually designed to prepare for undertaking further research on the subject. The literature review is often done a second time to help one interpret unexpected results. A *meta-analysis is a literature review in which the completeness of the survey is stressed, and statistical techniques are used to summarize the findings. Meta-analyses are sometimes called "systematic" reviews and are often contrasted with "traditional" or "narrative" reviews. Also called "research review."

LLR   *Log-likelihood ratio; often seen as −2LLR.

LN   Abbreviation for natural (not base 10) *logarithm, usually lowercase, ln.

Loading   See *factor loading.

Local Independence   In research on *latent variables, such as *factor analysis and *latent class analysis, local independence is the mathematical assumption that the latent variables account fully for the associations between the observed items. For example, in factor analysis all of the variance in the items is explained by the factor that includes the items.

Local Regression   Method for fitting a regression line or curve that makes no assumption about the shape of the line or curve. Rather, it follows the pattern of the data points, including any small (local) bends in and clusters of data points. One procedure for local regression is the *loess method.

Loess   Short for *lo*cally weighted regr*ess*ion. Methods for fitting smooth curves to a set of data. It uses *weighted least squares, which reduces the influence of extreme outliers. Also spelled "lowess," with the w standing for "weighted."

Lods   Short for *log odds, that is, logarithm of an *odds ratio. See *logit.

Log   See *logarithm.

Log$_e$ Symbol for natural *logarithm, *ln.

Logarithm   An *exponent of a number indicating the *power to which that number must be raised to produce another number.
    For example, the log of 100 is 2, because 10^2 (10×10) equals 100; the log of 1,000 is 3, because 10^3 ($10 \times 10 \times 10$) equals 1,000. The log of 47 is 1.6721 because $47 = 10^{1.6721}$. The "antilog" (or inverse log) turns the relation around; for example, antilog 2 = 100; antilog 3 = 1,000; antilog 1.6721 = 47. Logarithms were invented in the 17th century to ease the burden of calculations. Rather than multiply and divide large numbers, researchers could add and subtract their logs.
    When "log" or "logarithm" is used without qualification, this sometimes means "common logarithm," that is, logarithm using base 10, as in

the examples above. Statisticians more frequently use the "natural" (or "Napierian") logarithm (abbreviated *ln*), where the base is the *universal constant, *e* (2.71828.) The *ln* is widely used in *transformations to make data more closely resemble the *assumptions of statistical techniques, especially in *logistic regression. See *logit, *log-linear analysis.

Logic Model In *evaluation research, a graphic representation of the program elements (inputs) and their relationships that indicates how they will function to produce program outcomes. Logic models serve as frameworks for evaluation researchers as well as plans for program managers.

Logical Positivism The variety of *positivism most closely associated with the Vienna Circle of philosophers in the 1920s and 1930s and their followers. To the extent that "positivism" has any meaning beyond a vague and usually pejorative label used by qualitative researchers to describe narrowness and rigidity in scientific assumptions, it means the theories of logical positivists such as Rudolph Carnap and A. J. Ayer.

Logistic Model See *logit analysis/models.

Logistic Regression Analysis A kind of regression analysis often used when the *dependant variable is *dichotomous and scored 0, 1. (It can also be used when the dependent variable has more than two categories, in which case it is called "multinomial.") It is usually used for predicting whether something will happen or not, such as graduation, business failure, heart attack—anything that can be expressed as Event/Non-Event. Independent variables may be categorical or continuous in logistic regression analysis. *Ordinary least squares regression can be used when the *independent variables are dichotomous, but this is not good practice when the dependent variable is dichotomous.

Logistic regression is based on transforming data by taking their natural *logarithms so as to reduce nonlinearity. Only the *dependent variable is transformed; the independent variables are left in their *natural units. In other words, while linear regression uses the straight line that best approximates the data, logistic regression uses the logarithmic curve that best approximates it. Rather than using *OLS methods, logistic regression estimates parameters using *maximum likelihood estimation.

A logistic regression coefficient represents the effect of a one-unit change in an *independent variable on a dependent variable—specifically on the natural log of the odds of the dependent variable being in Category 1 when the dependent variable is *dummy coded (1, 0).

Logistic regression is an increasingly popular alternative to *discriminant analysis, because it requires fewer assumptions. Also called "logit regression analysis," although logit and logistic regression differ slightly. See *logit analysis/models.

L

Logit   Short for "logistic probability unit" or the natural "log of the odds." A *logistic regression analysis yields a probability of an event; that probability is transformed into an odds; the natural log of that odds is taken to get the logit. See *logit analysis/models. Compare *probit.

Logit Analysis/Models   A type of *log-linear analysis similar to multiple *regression analysis; it is used when both the independent variables and the dependent variable are *dummy (*dichotomous) variables. It is used for predicting a categorical dependent variable on the basis of two or more categorical independent variables. If one or more of the independent variables is continuous, then *logistic regression analysis is used. Compare *probit analysis.

   The terms "logit model" and "log-linear model" are sometimes used interchangeably, but it is more precise to say that the logit model is a particular type (probably the most commonly used type) of the log-linear model.

Log-Likelihood   The *logarithm of the *likelihood, which is more often used than the simple likelihood when conducting a maximum likelihood estimation.

Log-Likelihood Ratio (LLR)   A common *test statistic used for *logistic regression and *probit regression. The LLR is multiplied by -2. The test statistic is often written -2LLR or -2 LOG LR.

Log-Linear Analysis/Models   Methods for studying relations among *categorical (*nominal) variables in contingency tables. So called because they use equations that are transformed, by taking their natural logs, to make them linear. Log-linear analysis uses *odds rather than *proportions as is done in the more familiar *chi-squared tests. Log-linear models are capable of handling several nominal variables and their relations in a way that approximates *analysis of covariance structures. The results of a log-linear analysis can be tested either by the usual chi-squared *goodness-of-fit test or by the *likelihood ratio test.

   Log-linear models are an advance over the older *chi-squared test of independence, because unlike the chi-squared test, log-linear analysis can handle complex patterns of *interaction among the variables.

Log Odds   Another term for *logit, that is, the natural log of the odds. Note that *independence when using a log odds is zero, unlike for an odds, and in a logistic regression, where independence is 1.0.

Log-Rank Test   A statistical test for comparing sets of data in a *survival analysis. The technique involves comparing the ages of those who have survived to those who have not to see whether a *treatment has increased survival time.

Longitudinal Study (or Design)   A study over time of a variable or a group of subjects. See *panel study, *event history analysis, *prospective study. Contrast *cross-sectional study.

The National Educational Longitudinal Study of 8th graders starting in 1988 (NELS88) is a well-known example. Investigators began with large, national samples of students in the 8th grade whom they surveyed extensively. Every few years, the same students were contacted again to learn what they studied in high school and whether they graduated, whether and where they went to college, whether they graduated, what employment they had found, how much money they were making, and so on.

Lord's Paradox   A type of *Simpson's paradox that can occur when *change scores among preexisting groups are studied.

Lorenz Curve   A graphic representation of inequality or dispersion in a *frequency distribution. The cumulative percentages of a *population (such as income earners) is plotted against the cumulative percentage of another variable (such as percent of income earned). The more the curve departs from a straight line at a 45-degree angle, the greater the inequality. See *Gini coefficient, which is 2 times the distance between the curve and the diagonal line.

The following graph uses a Lorenz curve to plot the degree of inequality in Society A and Society B. Income inequality is much greater in B than in A.

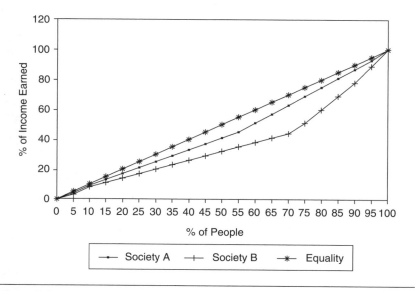

Figure L.2   Lorenz Curve

L

Lowess   See *loess.

Lowspread   The range of values in a *distribution between the *median and the lowest value. Compare *highspread, *box plot.

LR   *Likelihood Ratio.

Lurking Variable   A third variable that causes a *correlation between two others—sometimes, like the troll under the bridge, an unpleasant surprise when discovered. A lurking variable can be a source of a *spurious correlation. See also *confound. Compare *covariate, *latent, *mediating, and *moderator variables.

For example, if researchers found a correlation between individuals' college grades and their income later in life, they might wonder whether doing well in school increased income. It might; but good grades and high income could both be caused by a third (lurking or hidden variable), such as a tendency to work hard.

M Symbol sometimes used for the *mean (of a *sample).

Macro Prefix meaning "big." Used in economics and sociology to mean broad social, institutional, or system-level analyses or variables, such as the behavior of the economy as a whole. Usually contrasted with *micro, which refers to small-group or individual phenomena, such as individuals' job skills. Compare *molar.

MAD *Mean absolute deviation.

Mahalanobis Distance Abbreviated D^2. The difference (or "distance") between populations; it is used as a test of the assumption of homogeneity of variance. It is the most common test for the presence of multivariate *outliers.

Main Effect The simple effect of an *independent variable on a *dependent variable; the effect of an independent variable uninfluenced by (without controlling for the effects of) other variables. Used in contrast with the *interaction effect of two or more independent variables on a dependent variable. It is difficult to interpret main effects in the presence of interaction effects.

MANCOVA *Multivariate Analysis of Covariance, an extension of *ANCOVA to problems with multiple *dependent variables.

Manifest Function The obvious, ostensible, or purported use or purpose of a social phenomenon—usually contrasted with its *latent function. See *functionalism.

    For example, the manifest function of the death penalty might be deterrence. If you asked its supporters, this might be the reason (manifest function) they would offer for favoring it. But if there is little convincing

M

evidence that the death penalty deters crime, a functionalist might look for latent functions to explain the widespread support for capital punishment. Satisfying an urge for revenge might be the latent function.

Manifest Variable   A variable that can be directly observed. Often assumed to indicate the presence of a *latent variable. Also called an "indicator variable." See *factor analysis, *structural equation modeling.

For example, we cannot observe intelligence directly; it is a latent variable. But we can look at indicators such as size of vocabulary, success in one's occupation, IQ test score, the ability to play complicated games such as chess or bridge well, and so on.

Manipulated Variable   Another term for *independent variable. Also called "treatment variable." Compare *predictor variable. *Experimental researchers manipulate variables; *observational researchers do not.

Manipulation Check   Part of a pilot study to make sure an intervention or treatment is strong enough or has the intended effect. For example, if an experiment in social psychology was designed to measure the effects of subjects' anger on their perceptions, a manipulation check would ascertain whether in the pilot study the treatment succeeded in making subjects angry.

Mann-Whitney *U* Test   A test of the *statistical significance of differences between two groups. It is used when the *data for two *samples are measured on an *ordinal scale. It is a *nonparametric equivalent of the *t-test. Although ordinal measures are used with the Mann-Whitney test, an underlying continuous distribution is assumed. This test is also used instead of the t-test with interval-level data when researchers do not assume that the populations are normal. It is very similar to the *Wilcoxon test.

MANOVA   *Multivariate Analysis of Variance, which is an extension of *ANOVA to research problems with multiple dependent variables.

Marginal   (a) *noun* Totals of rows or columns in a *contingency table or *cross-tabulation. See *marginal frequency distributions. (b) *adjective* Small change or difference, often a change of one unit. See *marginal utility.

Marginal Distribution   See *marginal frequency distributions.

Marginal Effect   In *regression analysis, a marginal effect is the expected change in a *dependent variable given an infinitely small change in a continuous *independent variable.

Marginal Frequencies   See *marginal frequency distributions.

Marginal Frequency Distributions   Frequency distributions of grouped data in *cross-tabulations or *contingency tables. So called because they are found in the "margins" of the table. Often called "marginals" for short.

M

For example, researchers polled a sample of city residents about whether they favored busing to achieve school desegregation. They *cross-tabulated the answers by the race of the *respondents, with the following results. The totals are the marginal frequency distributions. The "row marginals" are 155 and 293; the "column marginals" are 187 and 261. These marginals could be used to calculate the *expected frequencies to use in a *chi-squared test of the *statistical significance of the findings. Doing so yields a chi-squared statistic of about 51, which at 1 *degree of freedom, is statistically significant ($p < .001$).

Table M.1   Marginal Frequency Distribution

		Black	White	*Total*
Favor Busing	Yes	103	52	155
	No	84	209	293
	Total	187	261	448

Marginally Significant   Just barely or almost significant. Sometimes used to describe research results that fail to exceed the *critical value needed to be *statistically significant, but that come close enough that the researcher wants to talk about them anyway.

Marginal Probability   A probability calculation that depends only on the frequencies in the margins of a table. See *expected frequency, *marginal frequency distribution.

For example, suppose you randomly sampled 1,000 adults in a large city and obtained from each of them their sex, height, and whether they were overweight. Your sample includes 600 men and 400 women. The marginal probabilities for sex would be .60/.40; for overweight they would be .30/.70, as in the bold numbers in the following table. The *expected frequency for each of the cells, if sex had no relation to being overweight, would be .18, .12, .42, and .28. If the actual, *observed frequencies departed significantly from those probabilities (expected frequencies) you could conclude that sex and being overweight are related in the population sampled.

Table M.2   Marginal Probability: Overweight by Sex, Expected Frequencies

	Men	*Women*	*Total*
Overweight	.18	.12	.30
Not Overweight	.42	.28	.70
Total	.60	.40	

M

Marginals   Short for *marginal frequencies, that is, the totals in a *cross tabulation. Row marginals are the totals for the rows; column marginals are the totals for the columns.

Marginal Totals   See *marginal frequency distributions.

Marginal Utility   The additional benefit that comes from obtaining a small (marginal) increase in some good, given the amount that you already have.

   For example, the benefit or utility of a glass of cold water on a hot summer day might be very great. The value to you of a second glass of water would probably be less great. After you drank the third or fourth glass of water, the marginal utility of one more would probably decline to almost nothing.

Margin of Error   A range of likely or allowable values. Although the term is widely used to report the results of survey research, in statistics the margin of error is usually expressed as a *confidence interval.

Markov Chain   *Time-series model in which an event's *probability is dependent only upon the immediately preceding event in the series, and this dependence is the same at all stages. The general idea is that the state of a system in the future will be unaffected by its past, except its immediate past. Also called "Markov process," "Markovian principle," and *chain path model.

   For example, if every year in a particular state 5% of city dwellers moved to the suburbs and 2% from the suburbs moved to the city, eventually the proportions would stabilize (with the same number moving each way) at around 29% city and 71% suburbs, no matter what the original proportions.

Markov Chain Monte Carlo (MCMC) Methods   Computer-intensive methods used for simulating large arrays of correlated observations. MCMC is a group of popular techniques for fitting complex models. Originally developed in physics, the techniques are now widely used in the social and behavioral sciences. The computer simulations are used to find probability distributions.

Matched Pairs   A *research design in which subjects are matched on characteristics that might affect their reaction to a *treatment. After the pairs are determined, one member of each pair is assigned at random to the group receiving treatment (*experimental group); the other group (*control group) does not receive treatment. Without random assignment, matching is not considered good research practice. Also called "subject matching."

   For example, if professors wanted to test the effectiveness of two different textbooks for an undergraduate statistics course, they might match the

M

students on quantitative aptitude scores before assigning them to classes using one or another of the texts. An alternative, if the professors had no control over class assignment, would be to treat the quantitative aptitude scores as a *covariate and control for it using an *ANCOVA design.

Matched Pairs *t*-Test   Synonym for *dependent samples *t*-test and *correlated samples *t*-test.

Matching   See *matched pairs.

Materialism   The philosophical position that physical matter is the only reality and that other sorts of phenomena, such as ideas and social values, are reducible to or are merely expressions of material reality. Compare *idealism, *empiricism.

Matrix   Any rectangular array of data in rows and columns. See *correlation matrix, *matrix algebra, *vector.

Matrix Algebra   Rules for adding, subtracting, multiplying, and dividing matrices. The advantage of matrix algebra is that it allows one to treat matrices as single objects. It is widely used in *regression analysis and other multivariate methods, because it greatly simplifies the calculations needed with studies containing more than two independent variables. See *vector.

The following example shows a very simple matrix operation, how to add two matrices, **A** + **B**, to get a third, **C**. (The letters symbolizing matrices are printed in boldface type.)

Table M.3   Matrix Algebra: Adding Two Matrices

	A	+	**B**	=	**C**	
4	8		4	3	8	11
9	6	+	2	4	= 11	10
1	2		8	6	9	8

Maturation Effect   A *threat to validity that occurs because of change in subjects over time. See *extraneous variable. Compare *history effect.

For example, to study the effects of a college education on social and political attitudes, we might ask entering students to complete an attitude survey. Three and one half years later, we could ask the same students (now seniors) to answer the same survey questions. Any changes might be due to

M

the effects of college, but they also might be due to the fact that the students have gotten older (matured) since we first surveyed them.

Mauchley's Test for Sphericity A test of the *assumption of *sphericity, which must pertain if *repeated-measures F tests of multivariate data are to be valid.

Maverick An *outlier so extreme that it is doubtful that it could belong to the *population being studied.

Maximin Strategy In *game theory, a strategy in which players try to maximize their minimum winnings or returns. Compare *minimax strategy.

An example might be concentrating one's investments in low-yield, but very safe, government bonds.

Maximum Likelihood Estimation (MLE) Statistical methods (usually alternatives to *OLS methods) for estimating the *population parameters most likely to have resulted in observed *sample data. MLE is an integral part of *SEM. It is also often used in *log-linear models to estimate *expected frequencies in *contingency tables.

OLS methods work by minimizing the sum of squared differences between observed and predicted scores. MLE chooses as the estimate of the parameter the value for which the probability of the observed scores is the highest. See *likelihood. The basic procedure in MLE is as follows: For each possible value a parameter might have, compute the probability that the particular sample statistic (observed values) would have occurred if it were the true value of the parameter. Then, for the estimate, pick the parameter for which the probability of the actual observation is greatest.

Maximum Likelihood Ratio Chi-Square (MLRCS) A test of the *statistical significance of *confirmatory factor analyses, *SEM models, and other statistical results that can be expressed as *likelihoods. Abbreviated L^2. The larger the values of this statistic, the poorer the fit of the model to the observed data. Unlike the ordinary (Pearson) chi-square, the MLRCS can be *partitioned.

Maxplane A method of *oblique rotation of the axes in *factor analysis.

MCA *Multiple classification analysis.

MCMC Methods *Markov Chain Monte Carlo methods.

McNemar's Chi-Squared Test A variation on the chi-squared test used when the samples are not *independent but are related in some way, as in before-and-after studies, and when the outcome variable is *dichotomous. See *correlated groups design, *Cochran's Q test.

MD Abbreviation for *median, and for *mean deviation.

Mean   (a) Average. To get the mean you add up the values for each case and divide the total by the number of cases. Often symbolized as M or as $\bar{x}$ ("X-bar"). For an example of how to calculate an arithmetic mean, see *mode. When used without specification, "mean" refers to the *arithmetic* mean. Much less commonly used are the *geometric* mean and the *harmonic* mean.

(b) The *geometric mean* is computed by taking the nth root of the product of n scores, e.g., the square root of 2 scores, the cube root of 3, and so on. For example, to get the geometric mean of 5, 7, and 9 you multiply $5 \times 7 \times 9 = 315$ and take the cube root to get 6.8, which is somewhat smaller than the arithmetic mean (7.0) for the same scores. The geometric mean is useful for such tasks as averaging *indexes.

(c) The *harmonic mean* of a series of numbers is calculated by dividing n by the sum of the *reciprocals of the numbers. For example, to get the harmonic mean of 40 and 60 you add 1/40 plus 1/60 and divide the result into 2, for a harmonic mean of 48. Like the geometric mean, the harmonic is always smaller than the arithmetic; and the harmonic is always less than the geometric. For instance, the harmonic mean of 5, 7, and 9 is 6.6.

Mean Absolute Deviation   An infrequently used measure of *dispersion. It is calculated using the *absolute values of the *deviation scores—not the squares of the deviation scores as is done when computing the *variance and *standard deviation. Also called *average deviation and "mean deviation."

Mean Square (MS)   Short for the mean of the squared *deviation scores, that is, the *variance. The *variance is most often referred to as the MS in an *ANOVA.

Mean Squared Error Criterion   A rule for choosing an *estimator designed to minimize *variance and *bias in the estimator. It is used when unbiased estimators may be less efficient than biased estimators, as when there is a problem with *multicollinearity.

Mean Square Residual (MSR)   Another term for *variance of estimate.

Mean Substitution   A procedure used when some values for *cases are missing. It involves replacing missing values with the mean of the values from the other cases. See *listwise deletion for an example.

Measurement   (a) Assigning numbers or symbols to things, usually to different characteristics of *variables. (b) The subdiscipline concerned with how to assign numbers or symbols to variables.

People often think of measurement and statistics as the same thing, but there is a distinction: Measurement is how we get the numbers upon which we then perform statistical operations. See *level of measurement.

Measurement Class   Another term for *class interval.

M

Measurement Error   Inaccuracy due to flaws in a measuring instrument or mistakes of those using it. Measurement error is inevitable, since perfect precision is impossible. If measurement errors are *random, they will cancel one another out in the long run. See *random error, *sampling error. Compare *bias.

For example, if a research team were studying the effects of stress on blood pressure, and the pressure gauge were not perfectly accurate (it never is), this would lead to measurement error.

Measure of Association   See *association, measure of.

Median   The middle score or measurement in a set of ranked scores or measurements; the point that divides a distribution into two equal halves. When the number of scores is even, there is no single middle score; in that case the median is found by taking the mean of the two middle scores. See the example at *mode.

Median Absolute Deviation   An estimate of variability spread of scores in a distribution. It is computed by subtracting the median of a distribution from each of the absolute values of the scores in a distribution and then taking the median of the resulting scores. Compare *mean absolute deviation.

Median Test   A *nonparametric test of significance to determine whether two *samples, measured on an *ordinal scale, come from populations having different medians. The *chi-squared test is applied to the proportions of the samples above or below the median. See *Mann-Whitney test.

Mediating Variable   Another term for *intervening variable, that is, a variable that "transmits" the effects of another variable. Compare *interaction effect, *moderator variable.

For example, parents transmit their social status to their children directly. But they also do so indirectly, through education, as in the following diagram, where the child's education is the mediating variable. See *path diagram.

Parents' Status → Child's Education → Child's Status

Figure M.1   Mediating Variable

MEDLINE   Abbreviation for Medical Literature Analysis Retrieval System onLine.

Megabyte   One thousand *kilobytes, or 1,024,000 *bytes.

Member Check (or Validation)   The practice of researchers submitting their data or findings to their *informants (members) in order to make sure they correctly represented what their informants told them. This is perhaps most often done with data, such as interview summaries; it is less often done with interpretations built on those data.

Memos A term used by some researchers to describe what is more traditionally called *field notes.

Meta-Analysis Quantitative procedures for summarizing or integrating the findings obtained from a *literature review on a topic. Meta-analysis is, strictly speaking, more a kind of *synthesis than analysis, and it is also called "research synthesis." The meta-analyst uses the results of individual research projects on the same topic (perhaps studies testing the same *hypothesis) as *data for a statistical study of the topic. The main controversies about meta-analysis have to do with identifying the appropriate studies to synthesize, that is, with specifying which studies are truly studying the same hypotheses, treatments, and populations.

Metaphysical Explanation An explanation not subject to physical (or observational or behavioral) testing. The term is most often used loosely by social and behavioral scientists to mean "unscientific," "hard to understand," and/or "highly unlikely."

Metatheory Theory about theory.
For example, a theoretical account of the *epistemological presuppositions of *logical positivism and *constructionism would be a metatheoretical work.

Methodological Individualism The *assumption that all generalizations about groups can be explained by (or reduced to) facts about individuals. Sometimes also called—more often by its opponents than its friends—*reductionism. Methodological individualism is often contrasted with *holism.
For example, take the statement, "Teachers' middle-class values often make them unable to respond to the needs of lower-class children." Methodological individualists would say that this statement makes no sense apart from the values of individual teachers and the needs of individual students; "middle-class" and "lower-class" are merely convenient generalizations, which when valid, summarize what we know about individuals.

Methodology (a) The study of research methods, from general problems bordering on *epistemology to specific comparisons of the details of various techniques. See *research design. (b) Sometimes a verbose way of saying "method," as in, "This article employs an interesting methodology." In this case, the difference between "method" and "methodology" is three syllables.

M

Method Variance   The effects (*bias) that different methods of measurement have on data collection—for example, written surveys versus face-to-face interviews. One argument for *multimethod and *mixed-method research is that by using more than one method, the biases of individual methods could cancel one another out.

Metric   Any standard or scale of measurement: inches, seconds, minutes, dollars, test scores, kilograms, and so on. The term is often used in a statement such as, "results are reported in the original metric," meaning that they have not been *transformed or *standardized.

Metric Variable   A variable that can be measured on an *interval or *ratio scale.

Micro   Prefix meaning "small." See *macro.

Micro-Data   Data about variables within a behavioral unit, such as an individual or a corporation. Micro-data is often contrasted with *aggregate data, which is about groups of behavioral units such as individuals grouped by race, sex, or class, or corporations grouped by economic sector.

Middle-Range Theory   Robert K. Merton's term for a theory describing relations at modest *levels of abstraction, or middling *levels of generality—somewhere between an *empirical generalization and a *metatheory. Middle-range theories are, as Goldilocks put it, "just right," not too atheoretical like empirical generalizations, but not too hard to test like metatheories.

Mid-mean   The mean of the *mid-spread, that is, the mean of the middle half of the values in a distribution.

Mid-range   The mean of the largest and smallest values in a sample.

Mid-spread   Another term for *interquartile range, that is, scores that range from the 25th through the 75th percentile.

Milgram Experiments   A series of studies of individuals' willingness to "just follow orders," even when doing so appeared to require hurting other people. The studies were controversial, both because of their shocking findings about how many people would be willing to obey evil orders, and because the methods used put the subjects of the study at risk of psychological harm. The experiments are often cited as an example of violating *research ethics. See *debriefing.

Minimax Strategy   In *game theory, a strategy in which players try to minimize their maximum losses. Compare *maximin strategy.

An example of where this strategy could be applicable might be designing power plants to avoid nuclear accidents. It might be wise to reduce the odds of meltdown (maximum loss) to the lowest possible level, even if that had to come at the cost of raising the odds of occasionally spewing small amounts of radioactive particles into the air.

MINITAB   A *software package for statistical analysis, often used for teaching statistics. Compare *SPSS and *SAS.

Missing at Random   When *missing data occur in ways that do not *bias samples, they are said to be missing at random. If the missing data have no systematic pattern (are randomly distributed), they cause only small problems for the researcher. In practice, it is often difficult to know whether missing data are missing at random.

Missing Data   Information not available for a subject (or case) about whom other information is available—as when a *respondent fails to answer one of the questions in a survey. For an example, see *listwise deletion.

Missing Values Procedures   Statistical methods for dealing with *missing data. See *listwise deletion, *pairwise deletion, *mean substitution, *multiple imputation.

Misspecification   An error in *regression analysis made by constructing a *model that excludes a *variable that ought to have been included—or includes one that ought to have been excluded. Compare *identification problem.

Mixed ANOVA   This form of ANOVA combines (mixes) *between-subjects factors, in which each subject is measured only once, and *within-subjects factors, in which subjects are measured more than once. In other words, subjects are measured only one time on some variables (e.g., sex), but more than once on others (e.g., scores on tests). Also called "split-plot" ANOVA. See *repeated-measures ANOVA, *mixed designs, definition (b).

Mixed Designs   (a) *Factorial designs in which the number of *levels of the factors is not the same for all factors. (b) Factorial *multiple regression analyses that combine *repeated measures (for the within-subjects variables) and one-time measures (for the between-subjects) variables.

Mixed-Effects Model   An ANOVA design combining the *random-effects model and the *fixed-effects model. Also called "Model III ANOVA."

Mixed Factorial Study or Experiment   Research in which the number of *levels of the *factors (or *independent variables) differs from one factor to another.

Mixed-Method Research   Inquiry that combines two or more methods. This particular term usually refers to mixing that crosses the quantitative-qualitative boundary. However, that boundary is not necessarily the most difficult one to cross. For example, mixing surveys and experiments (both quantitative methods) may require more effort for many researchers than

M

combining surveys and focus groups (the first quantitative and the second qualitative). Mixed-method research is often considered important for avoiding *method variance. See *triangulation, *multimethod research.

MLE   *Maximum likelihood estimate or estimation.

Mobility Table   A table showing persons' social or occupational status at two different times. Most commonly, individuals are cross-classified according to a parent's occupation (origin) and their own first occupation (destination).

Suppose we sampled 2,000 of the adult men in a large city and asked them two questions: When you were growing up and going to high school, what was your father's occupation? What was your first full-time job? If we assigned levels to the occupations and entered the results in a mobility table, it might look something like the following. The bold numbers on the diagonal are nonmobile sons, that is, sons who have the same rank as their fathers. Cells to the lower left of the diagonal show the numbers of upwardly mobile sons, while those to the upper right of the diagonal show the downwardly mobile.

Table M.4   Mobility Table: Father's Job by Son's First Job

		Son's First Job				
		Upper Middle	Upper	Middle	Lower Middle	Lower	*Total*
Father's Job	Upper	**120**	75	45	60	3	303
	Upper Middle	80	**50**	55	65	5	255
	Middle	80	65	**90**	170	10	415
	Lower Middle	70	90	90	**300**	45	595
	Lower	20	30	25	167	**190**	432
	Total	370	310	305	762	253	2000

Mode   The most common (most frequent) score in a set of scores. See *bimodal distribution.

For example, if students' scores on a midterm were distributed as in the following list, the mode would be 90, the *median 81, and the *mean 74. Be sure to avoid the mistake of confusing the modal category with the number of individuals in the category, the "modal frequency." The mode is the score or the category, in this case, 90; it is not the number of students earning that score, in this case 3.

Table M.5   Students' Midterm Scores: Mean, Median, and Mode

Student 1	94	
Student 2	90 ⎤	
Student 3	90 ⎬	→ the most common score (mode)
Student 4	90 ⎦	
Student 5	81 ⎤	→ the middle score (median)
Student 6	70	
Student 7	65	
Student 8	56	
Student 9	30	
Total	666	666 divided by 9 = 74 (mean)

Model   A representation or description of something (a phenomenon or set of relationships) that aids in understanding or studying it; a set of assumptions about relationships used to study their interactions, as a computer simulation might model economic developments, or as role-playing might be a model of social interaction. Compare *ideal type, *paradigm, *theory.

Usually the purpose of constructing a model is to test it. For an example of a graphic causal model, see *path diagram. Perhaps the most common form of model is an *equation, which is a model that states a theory in formal, symbolic language, as in a *regression equation.

Model I ANOVA   See *fixed-effects model.

Model II ANOVA   See *random-effects model.

Model III ANOVA   See *mixed-effects model.

Modeling   A term sometimes used to describe building a *model.

Moderating Effect   Another term for *interaction effect. Also called "conditioning" effect and "contingency" effect. Typically, the term moderating effect is used when the treatments or independent variables interact with *attributes of the people being studied, such as their age or sex. When two treatments or independent variables have a joint effect, this is more often called an interaction effect, not a moderating effect. Compare *effect modifier.

Moderating Variable   See *moderator (variable).

Moderator (Variable)   A variable that influences ("moderates") the relation between two other variables and thus produces a *moderating effect or an *interaction effect. Compare *mediating variable.

Modulus   (a) Another term for *absolute value. (b) The factor by which a *logarithm of a number in one base is multiplied to obtain the logarithm of the number in another base.

M

Modus Ponens   A rule of inference in the form: If A exists, then so does B; A exists, therefore B exists also. From the Latin meaning "method of affirming."

Modus Tolens   A rule of inference in the form: If A exists, then so does B; B doesn't, therefore neither does A. From the Latin meaning "method of denying." See *indirect proof.

Molar   Loosely, "big." Said of research concerned with whole systems or categories of subjects (large units of analysis), rather than the characteristics of the individuals making up the categories or systems. Usually contrasted with *molecular. See *macro.

Molecular   Loosely, "little." Having to do with parts rather than wholes, with simple rather than complex systems, with small rather than large units of analysis. Compare *molar. Molar and molecular are used more often in psychology than in sociology or economics. In the latter two disciplines a similar concept is captured by the *macro-*micro distinction.

Moment   The mean of the *deviation scores for a *variable that have been raised to a particular power. The *sum of squares is the most familiar example; its mean gives the *variance. Moments can be used for computing measures that describe a distribution; the first moment is used to calculate the *mean, the second the *variance, the third *skewness, and the fourth the *kurtosis of a distribution.

Monotonic Relation   Said of a relation between two variables in which an increase in one always ("monotonously") produces an increase (or decrease) in another. A monotonic relation is often, but not necessarily, a *linear relation; increases interrupted by periods of no change will still be monotonic as long as there is no reversal of direction. Compare *curvilinear relation, *nonmonotonic linear relation.

Monte Carlo Methods   Any generating of *random values (most often with a computer) in order to study statistical *models. Monte Carlo methods involve producing several sets of artificial data and using these to study the properties of *estimators. See *random variable. Compare *jackknife, *bootstrapping.

For example, statisticians who develop a new theory want to test it on data. They could collect real data, but it is much more cost-efficient, initially at least, to test the theory on sets of data (often hundreds of sets of data) generated by a Monte Carlo computer program.

Mortality   Another term for *attrition, or losing subjects in the course of a study, as in an experiment when some subjects withdraw. If those who opt out are different from those who stay (and there is no way to know for certain), this will compromise the *validity of the study.

Mortality Rate   The number of deaths occurring in a period of time (usually one year) as a *proportion of the number of persons in the *population.

Mortality Table   Another term for *life table.

Moving Average   In *time-series analysis, a method of *smoothing the line representing the data. Individual observations are replaced by a *mean of each observation and one or more observations on either side of it. By reducing the visibility of short-term *fluctuations, smoothing makes long-term *trends clearer. Two common kinds of moving average are the

M

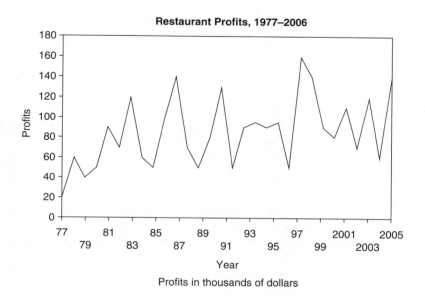

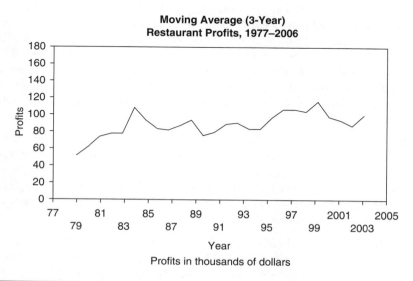

Figure M.2   Moving Average

three-year moving average (used in the above example) and the five-year weighted moving average. See also *ARIMA, an advanced smoothing technique.

MRA   *Multiple regression analysis.

MS   *Mean squares.

MSA   Short for SMSA, that is, *Standard Metropolitan Statistical Area.

MSR   Mean square residual. See *variance of estimate.

MTMM   Short for multitrait-multimethod models.

Mu [M, μ]   Lowercase mu is used to symbolize a *sample *mean.

Multicollinearity   In *multiple regression analysis, multicollinearity exists when two or more *independent variables are highly *correlated; this makes it difficult if not impossible to determine their separate effects on the *dependent variable. Also called *collinearity. See *intercorrelation, *tolerance.

Multidimensionality   Having more than one aspect or (dimension). Often used to describe attitudes.

For example, say a survey asked respondents for their overall attitude toward a presidential candidate. Any answers they give would likely mask the fact that their overall assessment was a composite of many dimensions, e.g., attitudes about the candidate's positions on foreign policy, welfare, and so on, as well as attitudes about his or her personal integrity, leadership qualities, and the like. In one sense, it is unrealistic to treat a complicated cluster of attitudes, some of which might be positive and others negative, as though it had only one dimension. In another sense, however, we often have to treat an "attitude object" on the basis of an overall attitude. In the case of a presidential candidate, this is especially clear. One must ultimately vote for or against—or not vote. The same kind of problem—a multidimensional issue requiring a monodimensional decision—can apply to many choices, such as whether to go back to school or to accept a job offer.

Multidimensional Scaling (MDS)   A method of using space on a graph to indicate statistical similarity and difference. Pairs of variables with the highest correlations are plotted closest together; those with the lowest correlations are furthest apart. MDS involves treating social or psychological distance as physical distance, using graphical distance in order to draw a map of how individuals' attitudes or characteristics cluster.

Multilevel Models   Models that include a hierarchy of *nested effects so as to disentangle the influences of different levels. For example, a multilevel study of the effects of schooling on academic achievement could study the effects of a particular classroom teacher (Level 1), who is located in a given school (Level 2), which is influenced by the policies of the district (Level 3) within which the school is located. See *hierarchical linear models.

Multimethod-Multitrait Models A version of *confirmatory factor analysis in which each factor (trait) is measured in several ways (methods) in order to reduce the distortion that any single measure always contains. Compare *mixed-method research and *triangulation.

Multimethod Research Another term for *mixed-method research, that is, research that combines two or more methods of *design, *measurement, or *analysis. Often a distinction is drawn between multi- and mixed-method research with mixed methods referring to joining methods that cross the quantitative-qualitative divide. Multimethod research is then used either to describe the joining of two or more quantitative methods or as a generic term for any combination of research methods. Whatever the label, using more than one method can be important for enabling the researcher to reduce biases likely to be associated with a single method. See *method variance, *triangulation, *convergent validity.

Multinomial Distribution A *probability distribution used to calculate the probabilities of distributions of discrete events that have more than two outcomes.

Suppose, for example, that at a certain college 40% of the students were freshmen, 30% sophomores, 20% juniors, and 10% seniors. Say we drew a random sample of 10 students (with replacement) and got two freshmen, three sophomores, five juniors, and zero seniors. The multinomial distribution could give us the probability of getting exactly that *sample distribution. That probability is, by the way, .0035.

Multi-operationalize To *operationalize a *construct in more than one way, that is, to measure a construct in more than one way.

For example, academic success of college students (the construct) could be measured by (operationalized as) grade point average, graduation, and/or rank in graduating class.

Multiple Classification Analysis (MCA) A technique used when the *independent (or *predictor) variables are "classificatory" (i.e., *nominal, *categorical, or *discrete) and the *dependent variable is measured on an *interval or *ratio scale. MCA results in *coefficients (etas and betas) that are *weighted according to the number of cases in each category of the independent variables. MCA is an alternative to using *dummy variables with regression analysis; it is often used to handle ANOVA designs in which there is an unequal number of cases in the cells.

Multiple Classification ANOVA Term for ANOVA designs using two or more *independent variables or *factors. Also called *factorial ANOVA. Contrast *one-way ANOVA.

M

Whatever the label, such designs allow the researcher to test for *interaction effects among the independent variables.

Multiple Comparisons   Making several comparisons from one set of data; most often discussed in regard to *ANOVA, where it is frequently called *post hoc comparisons. After obtaining a significant F statistic in an ANOVA, which tells the researcher that *some* means among treatment groups are significantly different, multiple comparisons involve looking among the possible comparisons between variables, trying to learn *which* differences are statistically significant. Several techniques exist for conducting such multiple comparisons while limiting the problem of inflated Type I error, for example, *Tukey's HSD test and the *Scheffé test. Compare *Bonferroni adjustment.

Multiple Correlation   A correlation with more than two *variables, one of which is *dependent, the others *independent. The object is to measure the combined influence of two or more independent variables on a dependent variable. R is the symbol for a multiple correlation coefficient. R^2 gives the proportion of the variance in the dependent variable that can be explained by the action of all of the independent variables taken together and is known as the coefficient of *determination.

For example, researchers could use multiple correlations to measure the combined effects of age and years of education on individuals' incomes.

Multiple Discriminant Analysis   See *discriminant analysis.

Multiple Imputation   *Missing value techniques that involve replacing each missing value with many values. This leads to many sets of "complete" data. These data sets are studied to better understand the nature and potential importance of the missing values.

Multiple Linear Regression   A method of *regression analysis that uses more than one *predictor variable (or *independent variable) to predict a single *criterion variable (or *dependent variable). The *coefficient for any particular predictor variable is an estimate of the effect of that variable while *holding constant the effects of the other predictor variables.

The generic term is "regression analysis," which when used without qualification means "*linear." "Multiple" means two or more independent variables, and most regression analyses are in fact multiple.

Multiple Regression Analysis (MRA)   Any of several related statistical methods for evaluating the effects of more than one *independent (or *predictor) variable on a *dependent (or *outcome) variable. Since MRA can handle all ANOVA problems (but the reverse is not true), some researchers prefer to use MRA exclusively. See *regression analysis. MRA answers two main questions: (1) What is the effect (as measured by a regression coefficient) on a dependent variable (DV) of a one-unit change in an independent variable (IV), while controlling for the effects of all the other IVs? (2) What is the total effect (as measured by the R^2) on the DV of all the IVs taken together?

Multiplicative Relations   Another term for *interaction effects; it is most often used when the research is *nonexperimental. So called because the effects of the variables are multiplied, not added.

Multiplier Effect   The tendency for an increase in investment or spending to have an effect that grows (multiplies) beyond the original amount spent or invested. The "multiplier" is a number expressing the extent of the multiplier effect. The term is used mainly in economics, but the concept of this kind of feedback effect has applications in other fields.

Multistage Sampling   Any sampling design that occurs in two or more successive steps or stages. The term is often used with *cluster sampling and *area sampling. Each stage increases the probability of *sampling error.

For example, suppose we wanted a sample of 3rd-grade students in U.S. schools. Because we do not have a list of all 3rd graders from which to draw our sample, we might do something like the following. First, Stage 1, we would take a (perhaps *stratified) sample from the roughly 15,000 school districts in the United States. Then, Stage 2, we could sample elementary schools within the districts chosen in the first stage; then, we could sample 3rd-grade classes within the schools; finally, students within the sampled classes would be sampled. This would be a four-stage cluster sample.

Multistrategy Research   Another term for *mixed-method research, specifically combining quantitative and qualitative methods.

Multitrait-Multimethod Matrix   A *correlation matrix used to examine the *convergent and *discriminant validity of a *construct. The matrix contains correlations between two or more constructs (traits) measured in two or more ways (methods).

Multivariate   Pertaining to three or more variables. In nearly all cases the prefixes "multi" and "multiple" can be used interchangeably.

Multivariate Analysis (Methods)   Any of several methods for examining multiple (three or more) *variables at the same time—usually two or more *independent variables and one *dependent variable. Usage varies. (a) Stricter usage reserves the term for *designs with two or more independent variables *and* two or more dependent variables. (b) More commonly, multivariate analysis applies to designs with more than one independent variable or more than one dependent variable or both.

Whichever usage you prefer, multivariate analyses allow researchers to examine the relation between two variables while simultaneously *controlling for how each of these may be influenced by other variables. Examples

M

include *path analysis, *factor analysis, *principal components analysis, *multiple regression analysis, *MANOVA, *MANCOVA, *structural equation modeling, *canonical correlations, and *discriminant analysis.

Multivariate Analysis of Covariance (MANCOVA)   An extension of *ANCOVA to research problems with multiple *dependent variables. See *MANOVA.

Multivariate Analysis of Variance (MANOVA)   The extension of *ANOVA techniques to studies with multiple *dependent variables. MANOVA allows the simultaneous study of two or more related *dependent variables while controlling for the correlations among them. If the dependent variables are not related, there is no point in doing a MANOVA; rather, separate ANOVAs for each (unrelated) dependent variable would be appropriate.

For example, to study the effects of exercise on at-rest heart rate, you could use ANOVA to test the (null) hypothesis that there is no difference in average heart rate of three groups: women who never exercise, who exercise sometimes, and who exercise frequently. MANOVA makes it possible to add related dependent variables to the design, such as mean blood pressure and respiratory rates of the three groups.

Multivariate Normality   Said of the values of variables when each variable in a multivariate analysis, and all linear combinations of those variables, is distributed normally. Multivariate normality is an *assumption of multivariate analyses, but one that can be difficult to test.

Mutually Exclusive   Said of two events, conditions, or variables when both cannot occur at once.

For example, subjects in a study cannot be both female and male, nor can they be both Protestant and Catholic, for those are mutually exclusive categories. However, they could be both female and Protestant because those are not mutually exclusive groups. Researchers using categorical variables should be certain that their categories are mutually exclusive—and *exhaustive.

N Number. Usage varies; among the most common meanings of the uppercase *N* are (a) number of subjects or cases in a particular study, (b) number of individuals in a population, (c) number of variables in a study.

n Number. Usage varies; among the most common meanings of the lowercase *n* are (a) number in a sample, as opposed to in a population, (b) number of cases in a subgroup.

For example, consider the following from a research report: "We interviewed a random sample of college graduates ($N = 520$) to get their opinions on several issues; males were 45% ($n = 234$) of the sample." This means that a total of 520 graduates were interviewed; 234 of them were in the male subgroup.

N! N *factorial. For example, 5! (5 factorial) means $5 \times 4 \times 3 \times 2 \times 1 = 120$.

NA Abbreviation for "Not Applicable" or "No Answer."

Napierian Log Another term for "natural log." See *logarithm.

Narrative Analysis Any of several approaches, mostly qualitative, to studying textual materials structured as a story, that is, an account of events held together by a common theme, usually including the passage of time. What was new about narrative analysis when it became important in the 1980s and 1990s was less its methods and more its subject matter. Narrative analysis made individuals' "stories" an important object of research.

Natural Experiment A study of a situation happening naturally, that is, without the researcher's manipulation, that approximates an experiment; variables occur naturally in such a way that they have some of the characteristics of *control and *experimental groups. The opposite of a natural experiment would be an "artificial experiment." Although this term is rarely used, it does capture the essence of the laboratory: an environment artificially purified of variables in which the researcher is not interested. Compare *natural setting, *observational research, *correlational research, *quasi-experiment.

For example, a solar eclipse provides astronomers opportunities to observe the sun that they cannot provide for themselves by experimental manipulation. Or, a comparison was once made between the number of dental cavities in a city with naturally occurring fluoridated water and a similar city without fluoridation. Or, comparisons of children's vocabulary growth during the school year versus during the summer months allow researchers to separate the effects of schooling from those of the children's home lives.

Naturalism   The belief that the social and human sciences should conduct their research with the same aims and methods as scientists who study other natural, but nonhuman, phenomena. Compare *positivism, *naturalistic observation.

Natural Logarithm   See *logarithm.

Naturalistic Observation (or Research)   Observation or research done in a *natural setting without any control or manipulation of the setting by the observer. The goal is to study subjects in their setting as they occur naturally, when they are not being studied.

Natural Setting   A research environment that would have existed had researchers never studied it. Used also to refer to behaviors and events that occur in those settings. See *natural experiment.

Among examples of social phenomena that seem to demand study by social scientists but that are not easy to put in a laboratory or otherwise manipulate are elections, unemployment, monetary inflation, riots, wars, poverty, kinship structures, marriage practices, and so on. These generally have to be studied in their natural settings, using observational rather than experimental techniques—or not studied at all.

Natural Units of Measurement   Units of measurement or scales that have not been *transformed or *standardized.

For example, saying that one year of experience leads on average to a $600 increase in income expresses the relation in natural units of measurement. By contrast, saying that a 1 *standard deviation increase in experience leads on average to a .23 standard deviation increase in income does not describe the relation in the units that people "naturally" use when discussing these variables. See *original metric.

N-by-M Design   Said of a *factorial research design in which each of the factors has more than two *levels. N and M stand for the number of levels of each factor, for example a 2×3 design.

NCE Abbreviation for *normal curve equivalent.

N-choose-K Short for the number of *samples of a particular size (K) that can be chosen from a given number (N) or *population of items. The terminology is often used in the explanation of *sampling distributions.

Necessary Condition In causal analysis, a *variable or event that must (necessarily) be present in order for another variable or event to occur. See *cause.

A necessary condition may or may not be *sufficient to produce an effect. For example, for it to snow, it is necessary that the air temperature be 32 degrees Fahrenheit or colder, but that is not sufficient; cold air is just one of the necessary conditions.

Negative Binomial Distribution In *probability theory, the *distribution of the number of *failures prior to the first *success (or other specific number of successes) in a sequence of *Bernoulli trials. Compare *Pascal distribution.

Negative Case Analysis A procedure used in *qualitative research for revising hypotheses. One begins with a hypothesis (about, say, the causes of urban riots) and systematically studies examples looking for disconfirming instances. As these are found, one revises the hypothesis in light of the negative evidence, resumes the search, and continues until no further disconfirming cases are found. Compare *indirect proof.

Negative Number A number that is less than zero; it is indicated by a minus sign in front of it; "-8" is minus or negative 8, as in 8 degrees below zero.

Negative Relation (or Correlation) Another term for *inverse relation, a relation between two *variables such that when one increases the other decreases and vice versa. Note that there is nothing "bad" or unfavorable about a negative relation; it is negative only in the sense that it is expressed by a negative number. For example, increasing the number of police patrols in an area might reduce its crime rate. Compare *positive or direct relationship.

Negative Results Said of a study that does not produce *statistically significant findings. This usage can be misleading, particularly when the lack of statistical significance can be a substantively ("positively") important discovery.

For example, a study that found no significant difference between groups that were believed to be different on some variable could make a positive contribution to our understanding of those groups. See *publication bias.

Nested Design (a) Said of a *factorial design in which *levels of one factor appear within only a single level of another factor. The opposite of such

N

"nested factors" are *crossed factors. (b) A synonym for *hierarchical models.

In the following examples (definition a), the speed of solving problems (*dependent variable) is studied as it is influenced by two factors: Levels of Difficulty (A) and Types of Reward (B). In both the crossed and the nested designs, there are three levels of difficulty. In the crossed design there are two types of reward; in the nested there are six. In the nested design, levels of reward B1 and B2 appear only in (are nested in) Level A1. In the crossed design, on the other hand, B1 and B2 appear at all three levels of A, but B3 through B6 do not appear at all.

Table N.1   Nested Design

Difficulty	A1		A2		A3	
Reward	B1	B2	B3	B4	B5	B6

Table N.2   Nested Design (Compared With Crossed)

Difficulty	A1		A2		A3	
Reward	B1	B2	B1	B2	B1	B2

Nested Variables   Said of variables located inside other variables—such as city, state, and national unemployment rates. See *multilevel models, *hierarchical linear models.

Net   Remaining after deductions. Net income, for example, is income after expenses have been deducted. Compare *gross.

Net of   After having *controlled for the effect(s) of some variable(s). Also called "net relationship."

For example, phrases such as the following often appear in research reports: "the influence of education level on political attitudes, net of the effects of age and region of residence . . ." This means: the influence of education on attitudes, having subtracted (*controlled for) any effects of respondents' age and where they live.

Network Analysis   Techniques for studying persons interacting in groups. Developed originally by anthropologists, it has become more widespread in political science, psychology, and sociology. The people interacting are studied in terms of the direction, frequency, duration, content, and so on of the interactions. The groups of interactors are studied in terms of their size, density, and so on. Sometimes called, especially in older works, *sociometry. See *directed graph.

N

Network Sampling   Another term for *snowball sampling.

Neural Net   Statistical and computer techniques used to try to replicate, and thus understand, the processes by which the brain learns.

Newman-Keuls Test (or Procedure)   A test for statistical significance used with multiple *post hoc comparisons. See *omnibus test, *Duncan test, *Tukey HSD.

Neyman-Pearson Theory   The general theory of *hypothesis testing, so named after its most important originators. See *Type I error, *Type II error, *power of a test.

Noise   In *information theory, any random disturbances to communication. The term originated from the analogy with static interfering with a radio transmission. This popular term is used broadly to refer to any *random error, such as *fluctuations around a *trend line in a *time series.

Nominal   In name only; for example, a researcher who talked of "a merely nominal distinction" would be referring to a difference in labels or names that concealed an underlying similarity. See *nominalism.

Nominalism   A philosophical doctrine to the effect that abstract *concepts (such as justice, virtue, or nothingness) are simply convenient labels or names; they do not refer to real entities. Compare *realism, *methodological individualism, *holism.

Nominal Scale (or Level of Measurement)   A scale of measurement in which numbers stand for names, but have no order or value. See *categorical variable.

   For example, coding female = 1 and male = 2 would be a nominal scale; females do not come first, two females do not add up to a male, and so on. The numbers are merely labels.

Nominal Variable   Another term for a *categorical (or a discrete or a qualitative) variable. See *nominal scale.

Nomographic   See *nomothetic.

Nomological Net   When a *concept or *construct is defined in terms of other concepts or constructs, those other concepts or constructs are its nomological net; they allow one to name it.

Nomothetic   Also called "nomographic." Said of research that attempts to establish general, universal, abstract principles or *laws. Also used to describe relations among variables as well as the research that tries to discover them. Nomothetic is often contrasted with *idiographic. Compare *etic, *emic.

Nonadditive   Said of a relation such that its total effect cannot be obtained by adding up its separate effects. See *additive.

For example, when there is an *interaction effect among the *independent variables in a study, the relation among those variables and with the *dependent variable is nonadditive.

Nondetermination, Coefficient of   That part of the *variance that cannot be explained by or accounted for by the measured effects of the *independent variable(s). Symbolized $1 - R^2$. Compare coefficient of *alienation, coefficient of *determination, *error term.

Nondirectional Hypothesis   Another (less accurate) term for a *two-direction hypothesis. See *directional hypothesis, *two-tailed test.

Nonexperimental Design   A research design in which the researcher observes or measures subjects without altering or controlling their situation. In experimental research, the investigator controls the *independent variables, and assigns subjects to treatments, but cannot do so in nonexperimental research. Also known as *correlational research, *observational research, and non-interactive research. Compare *experiment, *descriptive research, *quasi-experiment.

Nonlinearity   In *chaos theory and *catastrophe theory, nonlinearity refers to events that are not proportional to their causes, particularly when small causes lead to big events.

Nonlinear Equation (Model)   An equation that cannot be plotted on a graph as a straight line. Such an equation contains *powers higher than 1 and/or interaction effects, that is, the variables are combined by multiplication.

For example, $y = a + 2b$ is linear, but $y = a + b^2$ is not, nor is $y = a \times 2b$.

Nonlinear Regression   A regression problem in which the *parameters are not linear, which prevents the use of the *least squares criterion. See *polynomial regression, *spline regression.

Nonlinear Relationship   A relation between two variables, which, when plotted on a graph, does not form a straight line. See *curvilinear relationship for a graphic illustration.

Nonlinear Transformation   A *transformation of data such that when the original and the transformed data are plotted against one another on a graph, this does not result in a straight line. See *linear transformation.

The most common nonlinear transformations are done with logs, roots, and powers. These transformations change the relative distances between the data points in the original data.

The following graphs show a series of scores (2, 3, 4, etc.) that have been transformed in three ways. When they are multiplied times 10 (20, 30, 40,

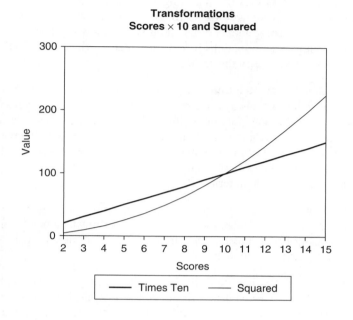

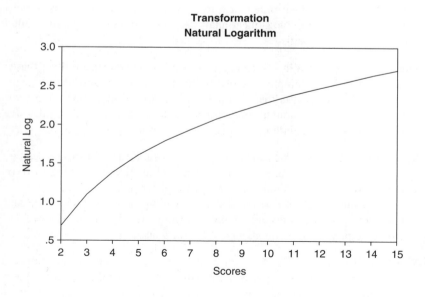

Figure N.1 Nonlinear Transformation

etc.) and plotted, this results in a straight line in the upper figure. When the original scores are squared (4, 9, 16, 25, etc.), this yields the curved line shown in the upper graph. A different curve occurs when doing one of the most common transformations in statistics, that is, when the natural log of the scores is taken, as in the second figure.

Nonmonotonic Linear Relation   A relation in which increases (or decreases) in one variable are always accompanied by increases (or decreases) in another, but the changes are not uniform. The direction of change is always the same (so the relation is linear), but the rate of change increases or decreases (so the relation is nonmonotonic). See *monotonic and *linear relations. Usage varies importantly. Many would consider "nonmonotonic linear" a contradiction in terms.

Nonorthogonal Designs   *Factorial designs are said to be nonorthogonal when the *cells, or *treatment groups, have unequal numbers; in other words, designs with unequal cell frequencies. Also called "unbalanced" designs. Compare *orthogonal.

Nonparametric Statistics   Statistical techniques designed to be used when the *data being analyzed depart from the distributions that can be analyzed with *parametric statistics. In practice, this most often means data measured on a *nominal or an *ordinal scale. Nonparametric tests generally have less *power than parametric tests. The *chi-squared test is a well-known example. See *log-linear analysis, *logistic regression. Also called *distribution-free statistics.

Non-Probability Sample   A residual category or label for everything that is not a *probability sample. The one thing all types of non-probability sample have in common is that it is not possible to estimate *sampling error when using them. Though often computed, tests of statistical significance are of limited value with non-probability samples.

Nonrecursive Model   A causal model which postulates that a variable can be both cause and effect, that there is a reciprocal relationship between two or more variables. See *simultaneous equations, *recursive model.
     For example, if you believed that education increases knowledge and that knowledge increases individuals' tendency to seek more education, you would be postulating a nonrecursive causal model.

Non-Response Bias   The kind of bias that occurs when some subjects choose not to respond to particular questions and when the non-responders are different in some way (they are a non-*random group) from those who do respond. See *missing values procedures.

N

For example, in a survey about taboo sexual practices, those who do not answer may be different from those who do; their missing information will bias the results.

Nonsense Correlation   A statistical association between variables that has no basis in reality, such as the correlation between the speed of data processing and the price of sunglasses over the past 20 years. See *spurious relation.

Non Sequitur   Latin for "it does not follow." An argument in which the conclusion does not follow from the premises.

NORC   National Opinion Research Center. Located at the University of Chicago, the organization is best known for conducting the *General Social Survey (GSS).

Norm   A standard of performance. In a *standardized test, the norm is determined by recording the scores of a large group, such as a sample of elementary school students. When subsequent students take the test, the norms (or standards) for them will be those of the larger group (that is, the group on which the test was "standardized"). Thus, for example, the expected *mean for subsequent students taking the test is the mean achieved by the original large sample of elementary students.

Normal Curve   See *normal distribution.

Normal Curve Equivalent (NCE)   A *standardized scale of scores developed by the U.S. Department of Education. Test takers scoring at the *mean get an NCE of 50, persons scoring in the 1st *percentile get a score of 1, and those in the 99th percentile get a score of 99. The *standard deviation for the NCE is 21.06. Compare *z-score.

Normal Distribution   A theoretical continuous probability distribution in which the horizontal axis represents all possible values of a variable and the vertical axis represents the probability of those values occurring. The scores on the variable (often expressed as *z-scores) are clustered around the *mean in a symmetrical, unimodal pattern known as the bell-shaped curve or normal curve. In a normal distribution the mean, *median, and *mode are all the same. There are many different normal distributions, one for every possible combination of mean and *standard deviation. Also sometimes called the "Gaussian distribution."

Since the *sampling distribution of a statistic tends to be a normal distribution, the normal distribution is widely used in *statistical inference. For small samples, the *Student's t distribution (which is also "bell-shaped" but not "normal") is preferable. See *region of rejection.

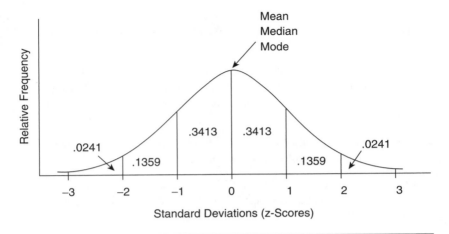

Figure N.2   Normal Distribution

Normalization   *Transforming the values of variables so that they approximate a normal distribution. This is done mostly so that statistical tests that *assume normally distributed variables can be used.

Normalized Standard Scores   The *standard scores of a distribution that have been *transformed so that they approximate a *normal distribution.

Normal Probability Plots   Graphic methods, using a *scatter plot, to examine the assumption that data come from a *normal distribution. If the sample is drawn from a normally distributed population, the points in the plot will approximate a straight line. This can be done by hand using normal probability paper, but it is more common to draw the plot with statistical software. The main types are the *detrend normal plot, *P-P plot, and *Q-Q plot.

The following figures all examine the same data on the salaries of 50 individuals. First the salaries are depicted in a *histogram. A *normal curve with the distribution's mean and standard deviation is superimposed on the histogram, which provides one way to judge the normality of the distribution. Next come two pairs of graphics: a P-P plot and its accompanying detrend plot, and a Q-Q plot and the detrend plot derived from it. The multitude of graphic and statistical tests for normality is an indication of how important the assumption of normality is for many statistical procedures.

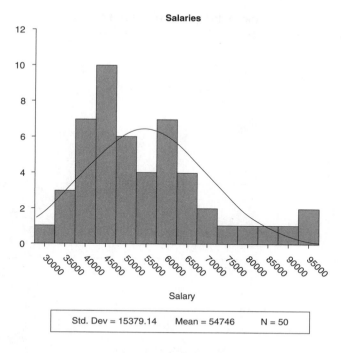

Salary

| Std. Dev = 15379.14 | Mean = 54746 | N = 50 |

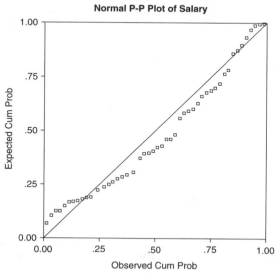

(Continued)

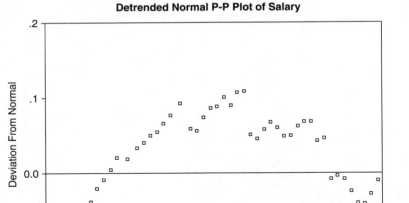

Detrended Normal P-P Plot of Salary

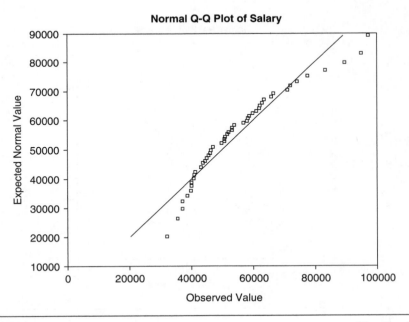

Normal Q-Q Plot of Salary

(Continued)

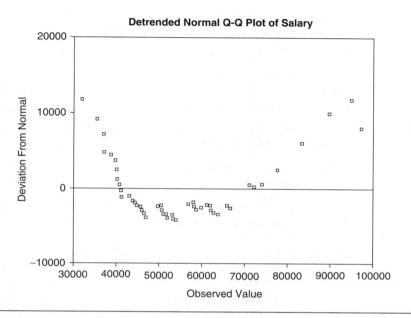

Figure N.3 Normal Probability Plots (P-P and Q-Q)

Normalization of Scores Converting original scores into scores on a standard scale. Often this is done by using the *percentiles of the scores and, assuming a *normal distribution, expressing these as deviations from the mean of a normal distribution. Thus the 50th percentile of the original scores would be treated as being at the mean, the 98th percentile would be treated as two *standard deviations above the mean, and so on.

Normative Pertaining to norms or standards. It is often used to refer to *prescribing* norms, standards, or values—as opposed to describing them.

Normative-Empirical Research Research undertaken with the goal of learning how to improve what is being studied. Compare *action research, *formative evaluation.

Normative Scale Generally any evaluative scale. Often used in contrast with *ipsative scale; see that entry for an illustration.

Norm-Referenced Test A test in which the scores are calculated on the basis of how subjects did in comparison to (relative to) others taking the test (others' scores provide the norm or standard). The alternative is some absolute standard or criterion. Compare *criterion-referenced test, *norm.

NS   Not significant (statistically significant, that is).

For example, the finding "$F = 2.38$, ns" means that the F ratio for that particular result was not large enough to reach *statistical significance.

Nu [N, ν]   A lowercase nu is often used to symbolize *degrees of freedom. $ν_1$ stands for the degrees of freedom in the numerator when computing an *F statistic; $ν_2$ stands for the denominator.

NUD*IST   Short for Non-numerical Unstructured Data Indexing, Searching, and Theorizing. A popular software for qualitative data coding and analysis.

Nuisance Variable   Another term for *extraneous variable. Compare *covariate, *lurking variable.

Null Hypothesis (H_0)   An hypothesis that a researcher usually hopes to reject, thereby substantiating its opposite. Often the hypothesis that two or more variables are not related or that two or more *statistics (e.g., means for two different groups) are not the same. The "null" does not necessarily refer to zero or no difference (although it usually does); rather it refers to the hypothesis to be nullified or rejected. In accumulating evidence that the null hypothesis is false, the researcher indirectly demonstrates that the variables are related or that the statistics are different. The null hypothesis is the core idea in *hypothesis testing. Compare *research hypothesis, *directional hypothesis, *falsificationism, *indirect proof.

The null hypothesis is something like the presumption of innocence in a trial; to find someone guilty, the jury has to reject the presumption of innocence. To continue with the analogy, they have to reject it beyond a reasonable doubt. The "reasonable doubt" in hypothesis testing is the *alpha level.

Null Set   In *set theory, an *empty set.

Numerator   In a fraction, the number above the line; the number into which the *denominator is divided.

Numerical Variable   Another term for *quantitative variable or *continuous variable.

Nuremberg Code   Standards for carrying out research on human subjects; written after the trials of Nazi war criminals following the Second World War revealed inhumane practices in medical experiments.

Objective   Said of a type of research or a finding that resembles an object in that it exists independently of the beliefs and desires of researchers or subjects. "Objective" is often used to refer to matters of fact rather than opinion. For example, people might disagree about whether a judge's ruling was fair, but they would be more likely to agree about the "objective fact" that she ruled against the plaintiff.

There should in principle be a high level of agreement about objective phenomena. But such consensus is fairly rare in the social sciences, largely because people often disagree over whether a particular sort of research or phenomenon is "really" objective. In practice, objectivity boils down to the level of consensus. If nearly everyone agrees that something is an objective fact, it becomes one more or less by definition. On the other hand, if there is much disagreement, it is hard to maintain that something is objectively true. This has led some writers to cease using the word "objectivity" and to replace it with *intersubjectivity, that is, consensus. Compare *subjective.

Objectivism   The belief that objective science is possible. The term is usually used by people who think objective science is *not* possible; they use it to describe the supposed beliefs of those who do.

Objectivity   (a) Those who think objectivity is possible often describe it as an approach that emphasizes fairness, accuracy, and lack of bias. (b) Those who think objectivity as defined in (a) is not possible often say that claims of objectivity are an attempt to disguise the self-interested conclusions of the powerful and privileged. See *relativism.

Oblique   Not at right angles; at an angle greater or less than 90 degrees. Used in *factor analysis and other research designs to refer to variables that are *correlated or not *independent of one another. In such cases in factor analyses, one uses oblique *rotation to allow the discovery of correlated factors. Compare *orthogonal.

Oblique Factor   A factor that is correlated with one or more other factors in a *factor analysis.

Observable Variable   A variable the values of which can be directly observed as contrasted with a *latent variable, the values of which have to be inferred. See *structural equation modeling.

Observational Research   Any of several research designs in which the investigators observe subjects but do not interact with them, as they would have to, for example, in interviews. Compare, however, *participant observation.

Usage varies. Some call almost any *nonexperimental research observational; others reserve the term for research by investigators who observe in a *natural setting, do not identify themselves as researchers, and do not participate in what they are observing. On the other hand, virtually all research is observational in one way or another; experimenters observe subjects as much as *participant observers do.

Observed Frequencies   When conducting a *chi-squared test, the term is used to describe the actual data in the *cross tabulation. Observed frequencies are compared with the *expected frequencies, that is, the frequencies you would expect if the *independent variable had no effect. Differences between the observed and expected frequencies suggest a relationship between the *variables being studied.

Observer Bias   Inaccuracies that occur when observers know the goals of the research or the hypotheses being tested and that knowledge influences their observations. Compare *blind analysis, *experimenter effect.

Observer Drift   The tendency, especially in lengthy research studies, for observers to become inconsistent in the criteria they use to make and record their observations. This causes a decline in the *reliability of the data they collect.

Ockham's Razor   A philosophical doctrine to the effect that theories and explanations should be as streamlined as possible. All other things equal, the simplest theory (for example, the explanation with the fewest predictors) is the best. Named after William of Ockham (1285–1349). The principle is more often called *parsimony today.

OCLC   Online Computer Library Center. A national bibliographic center in the United States for library materials containing, among other things, information about which libraries own which books. A kind of computerized national "card catalog."

Odds The *ratio of *success to *failure in *probability calculations.

For example, the odds of drawing, at random, a heart (success) from an ordinary deck of cards are 13 to 39, or 1 *to* 3, usually written 1:3. By contrast, the *probability (likelihood of success) of drawing a heart is .25, or 1 *out of* 4.

Odds Ratio (OR) A *ratio of one *odds to another. The odds ratio is a *measure of association, but unlike other measures of association, "1.0" means that there is no relationship between the variables. The size of any relationship is measured by the difference (in either direction) from 1.0. An odds ratio less than 1.0 indicates an *inverse or negative relation; an odds ratio greater than 1.0 indicates a *direct or positive relation. Also called "cross-product ratio" after a method of computing this statistic. See *logistic regression. See *risk ratio for an example comparing the odds ratio and the risk ratio.

An *adjusted* odds ratio is an OR computed after having controlled for the effects of other predictor variables. An unadjusted OR would be a *bivariate OR.

Official Statistics National census or survey data of the population collected by government agencies. Most nations assemble these data and make them available in statistical abstracts or other publications.

Ogive A graph of a *cumulative frequency distribution, so called because it resembles an architectural arch of the same name. Also called *sigmoid or S-shaped distribution.

OLS *Ordinary least squares.

OLS Regression The most common form of regression analysis; it uses the *least squares criterion for making estimates. When not otherwise specified, "regression" usually means OLS regression.

Omega Squared [Ω, ω] A measure of *strength of association, that is, of the proportion of the *variability in the *dependent variable associated with the variability in the *independent variable. Omega squared ranges from 0 to 1. When it is 0, knowing X (the *independent variable) tells us nothing at all about Y (the *dependent variable). When it is 1.0, knowing X lets us predict Y exactly. The omega squared for a particular study will yield an estimate smaller than either *eta squared or *R^2.

Omitted Variable Bias Mistakes made by leaving important causal variables out of an explanation. See *specification error, *misspecification.

Omnibus Test An overall test to determine whether there are any *statistically significant differences among three or more *treatment groups—such as the *F ratio used to test the results of an *analysis of variance. Omnibus tests are general; since they average all pairs of comparisons, they cannot specify what kinds of differences exist among which groups. See *planned comparisons, *post hoc comparisons.

One-Factor ANOVA   Another term for *one-way ANOVA.

One-Sided Test   Another term for *one-tailed test, that is, one in which the *alternative hypothesis is *directional.

One-Tailed Test of Significance   A *hypothesis test stated so that the chances of making a *Type I (or alpha) error are located entirely in one tail of a *probability distribution. Since a one-tailed test is less stringent than a *two-tailed test, it should be used only when there are very good reasons to make a *directional prediction. Also called "one-sided test." See *significance testing.

For example, suppose that in an experiment on the speed of vocabulary learning, the *null hypothesis is that there is no difference between males and females. A one-tailed hypothesis would be that one or the other sex learns faster. A two-tailed test would determine whether there was a statistically significant difference between the males and females studied, but it would not specify in advance what that difference was.

One-Way ANOVA   *Analysis of variance with only one *independent variable (IV) or *factor. Also called "single-classification" and "one-factor" ANOVA. Compare *factorial designs. Two-way ANOVA has two IVs, three-way has three IVs, and so on.

Open-Ended Question   See *open question format.

Open Question Format   A survey or interview format that allows respondents to answer questions as they choose. The questions are open-ended. Unlike a *closed question format, it does not provide a limited set of predefined answers.

Operational Definition   (a) A description of the way researchers will observe and measure a *variable; so called because it specifies the actions (operations) that will be taken to measure the variable. (b) The criteria used to identify a variable or condition.

Operational definitions are essential. They make *intersubjectivity (objectivity) possible because they can be replicated; operational definitions are always imperfect, usually by being artificial or too narrow. See *operations, *operationalize, *construct.

For example, the operational definition of an overweight person could be one whose *body mass index (BMI) is over 25. However, some highly muscled individuals have BMIs over 25 but would not fit most other definitions of overweight.

Operationalize   To define a *concept or *variable in such a way that it can be measured or identified (or "operated on"). When you operationalize a variable you answer the questions: How will I know it when I see it? How will I record or measure it?

For example, in a study of the academic achievement of poor schoolchildren, "poor" could be operationalized as eligibility for a subsidized lunch program, "achievement" as grade point average.

Operational Research   Another term for *operations research.

Operating System   In computers, the *program that controls the other programs and coordinates their interactions with one another and with *data files. Examples include DOS, OS, Windows, and Linux.

Operations   *Variables defined in such a way that they can be manipulated and measured. Such operations are ways to study more general *constructs or *theories. Also called *operational definition or "operationalizations."

For example, one of the ways the general construct of "job satisfaction" could be studied would be to use the absenteeism rate as a *proxy measure or operation of satisfaction (lower absenteeism rates would indicate higher satisfaction). See *construct, *operational definition, *operationalize.

Operations Research (OR)   A general approach to the scientific study of the activities (operations) of complex systems such as large corporations. Quantitative criteria are often used, and techniques are usually drawn from several disciplines, to make decisions about ways to increase the efficiency of the system as a whole. See *decision tree, *evaluation research.

Opportunity Cost   What one has to give up or postpone in order to do something else.

For example, deciding to use all your spare time to practice the piano means that you will not have any left to practice the violin. Or, going on for your master's degree might involve giving up an opportunity to take a good job.

OR   (a) *Odds ratio or (b) *operations research.

Oral History   Historical evidence and research based on memories of witnesses to events as these are reported in interviews, rather than memoirs or other written records.

Order   The number of variables controlled or held constant in studies of the relations between two variables. For example, a zero-order correlation is a correlation between two variables without introducing any controls. A first-order correlation controls for one, a second-order correlation for two, and so on.

O

Order (of a Matrix)   The number of rows and columns in a matrix; also called its "dimensions."

Order Effects   (a) In experiments where subjects receive more than one *treatment (a *within-subjects design), the influence of the order in which they receive those treatments. Order effects may *confound (make it difficult to distinguish) the treatment effects. To avoid this problem, experimenters often use *counterbalancing. See that entry for an example. (b) In survey research, when *respondents are asked more than one question about the same topic, the order in which the questions are asked can influence the answers. One example in which researchers experimented with question order involved asking U.S. respondents whether Russian newspaper reporters should be able to travel freely in the United States in order to gather information. Most respondents said no—unless they had first been asked whether U.S. reporters should be able to travel freely in Russia to gather information.

Order of Magnitude   The size of something expressed as a multiple of a number, usually in powers of 10.

Ordinal Interaction   Said of an *interaction effect that, when plotted on a graph, produces lines that do not intersect. Since the lines of any interaction effect are not parallel, those of an ordinal interaction *would* intersect if they were extended far enough, perhaps beyond the range of values of interest to the researcher. See *interaction effect and *disordinal interaction for a fuller definition and illustrations.

Ordinal Sampling   Another term for *systematic sampling.

Ordinal Scale (or Level of Measurement)   A way of measuring that ranks subjects (puts them in an order) on some variable. The differences between the ranks need not be equal (as they are in an *interval scale). Team standings or scores on an attitude scale (highly concerned, very concerned, concerned, etc.) are examples.

  A question that sometimes arises in statistical analyses is whether ordinal variables ought to be considered *continuous. A rule of thumb is that if there are many ranks, it is permissible to treat the variable as continuous, but such rules of thumb leave much room for disagreement.

Ordinal Variable   A variable that is measured using an *ordinal scale, such as the shirt sizes small, medium, large, and extra large. See *Kendall's tau, *Spearman's rho.

Ordinary Least Squares (OLS)   A statistical method of determining a *regression equation, that is, an equation that best represents the relationship between or among the *variables. See *least squares criterion. Compare *generalized least squares.

Ordinate   The vertical axis (or *y axis) on a graph. Compare *abscissa and see that entry for an illustration.

Organismic Variables   A term sometimes used to refer to *background variables.

Original Metric   Said of data that have not been transformed from the units of measurement used to gather them. See *natural units.

Original Research Report   A publication or other document based on primary data gathered by the researchers who wrote the report, not a publication based on somebody else's research. See *secondary source.

Orthogonal   (a) Intersecting or lying at right angles. (b) Used broadly to mean *independent or uncorrelated. Uncorrelated *variables are said to be orthogonal since, when plotted on a graph, they form right angles to one of the *axes (if there is no variance in one of the variables). More specifically, "at right angles" means "not correlated" because the cosine of the angle made by two lines is the *Pearson's correlation, and the cosine of a 90-degree angle is zero. (c) Research designs are called orthogonal if comparisons are between *independent groups, as when comparison groups have been determined by *random assignment. Compare *correlated groups. (d) In *factor analysis, said of a *rotation when the axes are kept at right angles, that is, when it is assumed that the factors are not correlated. Compare *oblique.

Orthogonal Coding   A method of coding in *regression analyses and ANOVAs. It is used to make *planned comparisons and test *hypotheses about the effects of *treatments on group *means. Also called "contrast coding." Compare *effect and *dummy coding.

Outcome Variable   Another term for *dependent variable, used mainly in nonexperimental research to refer to the presumed effect. Often used to describe the dependent variable when evaluating the effects of a treatment or an intervention. Also called *criterion variable and *response variable.

Outlier   A subject or other unit of analysis that has extreme values on a *variable. Outliers are important because they can distort the interpretation of *data or make misleading a statistic that summarizes values (such as a *mean). Outliers may also indicate that a sampling error has occurred by including a case from a population different from the *target population. See *skewed distribution, *trimmed mean, and *box-and-whisker diagram for a graphic illustration.

   For example, you might want to get the average (mean) income of households in your neighborhood so that you could argue that yours was not a rich neighborhood and it should not be subject to a tax hike. Your results might be something like those in the following table. The outlier is

Household 9. It raises the mean to $86,000, even though most households in the neighborhood do not earn even half that amount. To make your best case, you could either recompute the mean excluding Household 9 (which would give you a figure of $36,250). Or you could use the *median income ($38,000). Since the median is more *resistant to the effects of outliers, it is more accurate than the mean as a measure of *central tendency for your neighborhood.

Table O.1   Outlier: Neighborhood Household Income

Household 1	$22,000	
Household 2	$27,500	
Household 3	$28,000	
Household 4	$35,000	
Household 5	$38,000	(median)
Household 6	$40,000	
Household 7	$49,000	
Household 8	$50,500	
Household 9	$484,000	
Total	$774,000	774,000/9 = $86,000 (mean)

Outlying Case   Another term for *outlier.

Overall Regression Equation   A *regression equation in which terms for *interaction effects are included, that is, in which *product variables are calculated and product terms included.

Overidentified Model   In *regression analysis, a model containing more information than necessary to estimate regression coefficients. Compare *underidentified model and *just-identified model.

Overparameterized Model   A model with more *parameters to be estimated than observations with which to make the observations. Comparisons among states in the United States have a maximum *N of 50, but there are many more than 50 variables about states that one might wish to estimate.

Oversampling   A technique in *stratified sampling in which the researcher selects a disproportionately large number of subjects from a particular group (stratum). Most often researchers oversample in a stratum that would yield too few subjects if a simple *random sample were used. For example, if one were to conduct a survey to compare the attitudes of male and female nurses, simple random sampling might select too few males for analysis.

P (upper- and lowercase *P*, usage varies considerably) (a) Symbol for sample proportion, that is, the frequency of a particular event divided by the size of the sample. For example, if the sample were 20 coin flips, 9 of which came up heads, the sample proportion for heads would be .45 (9 heads/20 flips), $P = .45$.

(b) In path analysis, the symbol for a path, usually written with subscripts indicating the particular path and the direction of causal influence. For example, p_{32} means the effect of variable 2 on variable 3.

(c) *Probability value, or *p value. Usually found in an expression such as $p < .05$. This expression means, "the probability (p) that a result this big could have been produced by chance (or *random error) is less than ($<$) five percent (.05)." Thus, the smaller the number, the greater the likelihood that the result expressed was not merely due to chance. For example, $p < .001$ means that the chances are one in a thousand (one-tenth of 1%) against the result being a fluke. What is being reported (.05, .001, and so on) is an *alpha level or *significance level. The p value is the actual probability associated with an obtained statistical result; this is then compared with the alpha level to see whether that value is (statistically) significant, that is, whether it is less than the alpha level. For example, $p = .04$ is usually reported as $p < .05$.

P-P Plot Probability-Probability plot. A graphic method of comparing two *probability distributions by plotting one against the other, often for comparing a normal distribution with sample data to determine whether the sample data are normally distributed. Compare *Q-Q Plot. See *normal probability plot for a graphic illustration; see also *probability paper, *probability plots.

P

Paired *t*-Test   A *t*-test used to compare samples that are *correlated, such as scores of the same subjects on a *pretest and a *posttest. The opposite is an *independent samples *t*-test. Also called a correlated groups *t*-test.

For example, a program designed to raise achievement scores is implemented. After the conclusion of the program, to see whether the means of the pretest and posttest scores were significantly different, the *t*-test for pairs should be used.

Pairs   When referring to all possible pairs of observations in a *sample, a way of arranging the *data so as to be able to use one of several *measures of association for *ordinal variables such as *Kruskal-Wallis or *Mann-Whitney U.

Pairwise   Two at a time, as when the *means of groups are compared in pairs, two at a time.

Pairwise Comparison   Comparing two (not more than two) individual or group scores. Compare *omnibus test.

Pairwise Deletion   Removing a case from the calculation of a *correlation coefficient (or other statistic) when it has missing values for one of the *variables. The pairwise procedure does not reject the case altogether, but only sets it aside for the pair of variables for which information is missing. Compare *listwise deletion, and see that entry for an example.

Panel Study   A *longitudinal study of the same group (or "panel") of subjects. Compare *cohort analysis, *cross-sectional study. A panel study surveys the same individuals at different times, while a cohort study samples from the same group at different times. Because of this difference, a panel study is sometimes thought of as a true longitudinal study; a cohort study is an approximation.

Paradigm   A discipline's general orientation or way of seeing its subject matter. This meaning, today the most common in the social and behavioral sciences, was introduced by Thomas Kuhn. Originally "paradigm" referred to an example in grammar showing a pattern in a conjugation or declension (e.g., ring, rang, rung; sing, sang, sung). Compare *schema, *model.

Physics around the time of Einstein is said to have undergone a "paradigm shift"—from one understanding of the discipline and the world it studied to a radically different one. Fields such as political science and sociology are sometimes referred to as "multi-paradigm" disciplines, since there are several competing ways of understanding those disciplines and their problems.

Paradox of Inquiry   A puzzle in research that can arise when one wants to study an unknown subject. In the *Meno*, Socrates is asked (roughly): How can we seek something if we don't know it, and if we already know it, why would we seek it?

This paradox has important parallels to analytical problems in the social and behavioral sciences, most clearly to *specification error in *regression and *path analysis. If you do not have the "right" model, you cannot measure the effects of variables, but there is no way to know if you have the right model apart from the measured effects of variables.

Parallel Distributed Processing   Information processing involving a large number of units working simultaneously. Refers both to how the brain is thought to work and to how some computer networks do.

Parameter   Most broadly, a parameter is either (a) a limit or boundary or (b) a characteristic or an element. The word has many general and technical uses.

In statistics, a common use of "parameter" is for a characteristic of a *population, or of a distribution of scores, described by a *statistic such as a *mean or a *standard deviation. For example, the mean (average) score on the midterm exam in Psychology 201 is a parameter. It describes the population composed of all those who took the exam. Population parameters are usually symbolized by Greek letters, such as sigma, not Roman letters such as *s*, which are used for sample statistics.

In computers, the most common use of "parameter" is as an instruction limiting or specifying what you want the computer to do. For example, if you typed the following into your computer: "delete files 4, 7, & 9," "delete" would be the command; "files 4, 7, & 9" would be the parameters.

In mathematics, "parameter" means an unknown value that may vary. In the equation $Y = bX + e$, the parameters are b and e.

Parameter Estimation   Inferring a *population characteristic (parameter) from information about a *sample. There are as many methods of parameter estimation as there are kinds of *inferential statistics.

Parametric Statistics   Statistical techniques designed for use when *data have certain characteristics—usually when they approximate a *normal distribution and are measurable with *interval or *ratio scales. Also, statistics used to test hypotheses about *population parameters. Compare *nonparametric statistics or *distribution-free statistics.

Parsimony   Generally, frugality or thriftiness. Used in methodological writing to mean a principle for choosing among explanations, theories, models, or equations: The simpler the better, less is more. Of course, applying this standard makes the most sense when the explanations one is choosing among are about equally good except for their degree of simplicity. See *Ockham's razor.

For example, the smaller the number of predictor variables used to predict an outcome variable, the more economical or parsimonious the prediction equation.

P

Part Correlation Another term for *semipartial correlation, that is, a multiple
correlation in which a variable is partialed out, but only from one of the
other variables. Not to be confused with *partial correlation.

Partial Correlation Called "partial" for short. A correlation between two
*variables after the researcher statistically subtracts or removes (*controls
for, *holds constant, or "partials out") the linear effect of one or more other
variables. The opposite of a partial relation is not a "whole" relation, but
rather a simple relation, that is, one uncomplicated by considering other
variables. The difference between a partial correlation and a *beta weight
is that the beta weight is an *asymmetric measure, while the partial is
*symmetric. See *covariate, *control variable, *semipartial correlation.

  Symbolized r with subscripts. For example, $r_{12.3}$ means the correlation
between variables 1 and 2 when variable 3 is controlled; $r_{13.2}$ means the
correlation between 1 and 3 when 2 is controlled. Another way to put it:
$r_{12.3} = .27$ means that the correlation between variables 1 and 2 would have
been .27 had all the subjects been alike with respect to variable 3.

Partial Least Squares Regression A merging of the techniques of multiple
regression with those of *principal components analysis that is especially
useful when researchers want to predict dependent variables from a very
large number of independent variables.

Partial Out To *control for or *hold constant. See *partial correlation, *partial relations.

Partial Regression Coefficient A *regression coefficient in a *multiple
regression equation. So called because each regression coefficient in a multiple regression equation shows the part of the variance in the dependent
variable associated with one of several independent variables. The partial
regression coefficient estimates the difference in the dependent variable
associated with a one-unit difference in an independent variable—when
"partialing out" (*controlling for) the effects of the other independent variables. Also known as the (partial) regression weight and the (partial) slope
coefficient.

Partial Relations Called "partials" for short. (a) Relations among variables
discovered by dividing a sample into "parts" or subsets in order to test for
(or *control for) the effects of additional variables. (b) More generally, relations between two variables with one or more other variables controlled,
whether or not they are controlled by the partial table technique.

  For example, say we did a survey and found that young adults (aged
20–39) were less likely to be racially prejudiced than older adults (aged
40–59). We thought that the differences might be due not only to age but
also to the fact that the older adults in our sample tended to be less educated
than younger adults. We could see if that were true by "partialing," that is,

we could divide the total sample into a more educated group (13+ years of schooling) and a less educated group (12 years or fewer). We could then recompute the level of prejudice and the effects of age in each of the two parts of the sample. By comparing the *partial tables produced in this way, we could get a better understanding of the partial relationships between education and prejudice, between age and prejudice, and between education and age.

Partials   Short for *partial relations and *partial correlations.

Partial Table   A subtable of *cross tabulations for two *variables formed on the basis of the outcomes on a third (*control) variable. See *partial relations.
   For example, if we computed a cross tabulation table for coffee drinking and heart disease, we might want to *control for sex by constructing separate sub- or partial tables for men and women.

Participant   Another term for a research *subject. The use of "participant" arose apparently because many found the use of "subjects" for the persons being studied to be offensive. See *informant, *respondent.

Participant Observation   A kind of investigation in which a researcher participates as a member of the group she or he is studying. Sometimes the researcher informs the group that she or he is an observer as well as a participant, and sometimes the researcher pretends to be an ordinary member. Ethical dilemmas most often arise in the latter case, that is, when researching "under cover." See *ethnographic research.

Participatory Evaluation   Approaches to program *evaluation research that include program staff and clients in the process of evaluation, often as part of a *formative evaluation. The conceptual opposite of participatory evaluation would be an external audit for purposes of *summative evaluation.

Participatory Research   Another term for *participant observation and one that is associated more with sociology than anthropology. Compare *action research.

Partition   (a) *noun* Two or more subdivisions or categories of a *factor. For example, if class is the factor, upper, middle, and lower could be partitions. (b) *verb* In *set theory, to divide a *universal set into subsets that do not intersect and that exhaust all the elements in the universal set. (c) *verb* In *analysis of variance, to break the total *variance of observations into parts for purposes of *analysis. The variance is partitioned into two parts: an explained part, which is due to *regression or to *between-groups differences, and an unexplained part (called *error or *residual variance), which comes from differences among subjects*within groups. (d) *verb* In *multiple regression, to divide the explained variance (R^2) into parts accounted for by different independent variables or groups of independent variables.

P

PASCAL   A programming language. Like BASIC, C, COBOL, FORTRAN, and others, it is used to write computer programs.

Pascal Distribution   A *probability distribution used to calculate the number of *trials necessary to get a particular number of *successes. Also called *negative binomial distribution. Compare *binomial distribution, *geometric distribution.

For example, we would use a Pascal distribution if we were interested in the number of flips of a fair coin (trials) it would take to get a total of 10 heads. The probability of getting 10 heads in just 10 or 11 flips would be very small; so would the probability of needing 40 or more flips to get 10 heads.

Path Analysis   A kind of *multivariate analysis in which causal relations among several *variables are represented by graphs (*path diagrams) showing the "paths" along which causal influences travel. The causal relationships must be stipulated by the researcher. They cannot be calculated by a computer; the computer is used to calculate *path coefficients, which provide estimates of the strength of the relationships in the researcher's hypothesized causal system. Path analysis is an early form of *structural equation modeling.

In path analysis, researchers use data to examine the accuracy of causal models. A big advantage of path analysis is that the researcher can calculate direct and indirect effects of independent variables.

Path Coefficient   A numerical representation of the strength of the relations between pairs of *variables in a *path analysis when all the other variables are held constant. Path coefficients are *standardized regression coefficients (*beta weights), that is, they are regression coefficients expressed as *z-scores. Unstandardized path coefficients are usually called *path regression coefficients. See *path diagram.

Path Diagram   A graphic representation of a hypothesized causal model. Also called "flow graph."

For example, the following path diagram shows the effects of educational and other *background variables on occupational attainment. Subjects' job status, the *dependent variable, is determined by three variables (parents' education, parents' job status, and the subjects' own education).

How to Read a Path Diagram

1. We can see that parents' education and job status affect their children's job status directly; they also do so indirectly, by way of their influence on children's education.

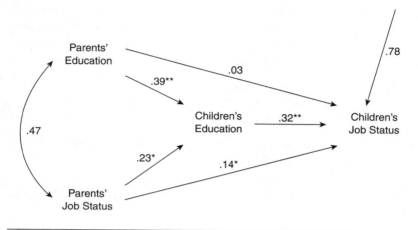

Figure P.1   Path Diagram

NOTE: *p < .05; **p < .01

2. The curved, two-headed line indicates that while there is a relation between parents' job status and education (.47), no causal assumptions are made about it.

3. The numbers on the lines are beta weights (*path coefficients). These are expressed in *z-scores (standard deviation units). For example, the .32 on the line between children's education and job status means that every 1.0 standard deviation increase in education level leads to a .32 standard deviation increase in children's job status. If the path coefficients were negative (none in this diagram are) that would indicate an *inverse (or *negative) relation.

4. The figure .78, on the line coming from outside the system, is the *residual (often abbreviated "e" or "u"). It is the part of the *dependent variable not explained by the *independent variables. To find the percentage unexplained by the total model, you need to square the residual coefficient: .78 × .78 = .608, which means that 60.8% of the total variance in children's job status is unexplained by the model pictured in this diagram.

5. Indirect effects can be computed by multiplying path coefficients. For example, to get the indirect effect of parents' education on children's job status, you multiply the effect of parents' education on

children's education (.39) times the effect of children's education on children's job status (.32). This gives you an indirect effect of .12 ($.39 \times .32 = .1248$), which is considerably bigger than the direct effect (.03) of parents' education on children's job status.

6. The asterisks (*, **) indicate the *statistical significance of the path coefficients. For example, .23* means that coefficient is significant at the .05 level, whereas the .39** is significant at the .01 level.

Path Regression Coefficient   Another term for an unstandardized regression coefficient in a *path analysis. See *path coefficient.

Pattern Recognition   Methods of classification that allow researchers to discover patterns among the values of variables, most importantly *cluster analysis and *discriminant analysis.

Pattern Variable   A *nominal or *categorical variable whose categories are made of combinations ("patterns") of other nominal variables. Pattern variables are often used to study *interaction effects. See *contingency table.

For example, if we know the employment status (employed/unemployed) and the sex (male/female) of a group of subjects, we could have four pattern variables: employed women; unemployed women; employed men; unemployed men.

PC   Percent correct. Not to be confused with politically correct (or private corporation or prince consort).

PCA   *Principal components analysis.

PDI   *Percentage Difference Index.

PDF   (a) *Probability Density Function. (b) Portable Document Format, usually lowercase pdf. Software for sending documents on the Internet; it allows one to make documents available for downloading and printing, but prevents them from being modified.

Peacock Effect   The tendency of males in a study, such as a social psychology experiment or a focus group, to show off for the females.

Pearson's Chi-Squared Statistic   What is usually meant when "chi-squared" is used without qualification. See *chi-squared test. Compare *likelihood ratio chi-squared.

Pearson's Contingency Coefficient   A measure of association that can be used when both the *dependent and *independent variables are *categorical. Symbolized C. It is based on the *chi-squared statistic and was originally an extension of the *phi coefficient to tables with more than four *cells. See *Cramer's V.

Pearson's Correlation Coefficient   More fully, the Pearson product-moment correlation coefficient. More briefly, Pearson's *r*.

A statistic, usually symbolized as *r*, showing the degree of *linear relationship between two *variables that have been measured on *interval or *ratio scales, such as the relationship between height in inches and weight in pounds. It is called "product-moment" because it is calculated by multiplying the *z-scores of two variables by one another to get their "product" and then calculating the average (mean value), which is called a "moment," of these products. Note: Pearson's *r* is rarely computed this way; the preceding is known as the "definitional," not the "computational," formula. See *correlation coefficient.

Pearson's correlation is so frequently used that it is often assumed that the word "correlation" by itself refers to it; other kinds of correlation, such as *Kendall's and *Spearman's, have to be specified by name.

Correlation and regression are often discussed together. This is because correlation is a special case of regression. Pearson's *r* is a *standardized regression coefficient (or *beta) between two variables. The essential link between the two is most easily discussed by referring to a scatter diagram that includes a regression line. The regression line is the line that comes closest to the points on the diagram. Correlation is the degree to which the points come close to the line. If the correlation were perfect (−1.0 or +1.0) all points would be on the line.

In the following example, showing the association of education levels and birth rates in 100 countries, the points definitely are arranged in a

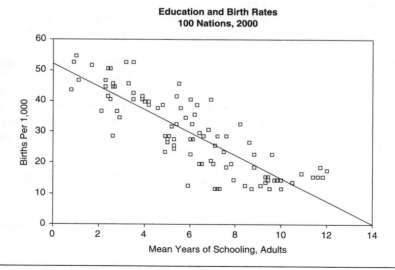

Education and Birth Rates
100 Nations, 2000

Figure P.2   Pearson's Correlation Coefficient

linear pattern. Since the line and the pattern of points run from the upper left to lower right, the correlation is negative, as is the regression coefficient. The correlation coefficient is −.84. This is also the standardized regression coefficient. The unstandardized regression coefficient is −3.7, which means that, on average, for every one-year increase in education, the birth rate goes down by 3.7 per 1,000. See *intercept for another example.

Pearson's *r*   See *Pearson's correlation coefficient.

Percent   Per hundred; one part in one hundred; one part of a whole that has been divided into 100 parts. If you multiply a *proportion times 100, you get a percentage.

Percentage Difference Index (PDI)   An index calculated by subtracting one percentage from another.
   For example, if 30% of the voters in your town wanted to replace the property tax with a sales tax and 70% were opposed, the PDI would be −40 (30 − 70 = −40).

Percentage Frequency Distribution   A *frequency distribution that shows the percentage (not the number) of cases having each of the *attributes of a particular *variable. See *relative frequency distribution.
   For example, say that in Economics 101, 22% of the students got A's, 36% B's, 29% C's, 11% D's, and 2% E's. The variable is grade; the attributes are A, B, C, D, and E; and the frequencies are 22%, 36%, 29%, 11%, and 2%.

Percentaging Rule   This rule says that in cross tabulations, percentages should be calculated within the categories of the *independent (not the *dependent) variable.

Percentile Rank   A number or score indicating rank by telling what percentage of those being measured fell below that particular score.
   For example, if you scored at the 74th percentile on some part of the Graduate Record Examination, this means that your score cuts off the bottom 74% of the distribution; your score exceeded that of 74% of the others who took the test (or, sometimes, 74% of those upon whom the test was *standardized).

Period Effects   Influences on people of a particular era or time (period). Compare *age effects, *cohort effects.
   For example, living in a period in which the presidency has been held by political conservatives for two decades may have influenced Americans' attitudes toward their government.

Periodicity   Said of something that recurs regularly, that is, that has periods.

Permutation (a) The process of changing the order of an ordered set of objects. (b) An ordered sequence of *elements from a *set.

A permutation is often contrasted with a *combination*. AB and BA are two different permutations of A and B, but they are the same combination. The mixed doubles team of Jane and Dick is the same team (combination) as Dick and Jane. But if Jane is first in a contest and Dick is second, that is a different result (permutation) than if Dick is first and Jane is second. Note that for any set of two or more objects, the number of permutations is greater than the number of combinations, because many permutations are just reorderings of a single combination.

Permutation Tests A category of *distribution-free tests of statistical significance, especially tests of rank-ordered data. They use techniques similar to the *bootstrap, but rearrange ("permute") the sample rather than sample with replacement.

Personal Probability A person's judgment or degree of belief that something will occur. This is contrasted with the method of estimating probability that is built on the empirical frequency of an event in the past.

Person Time Total time devoted, summed over persons, such as 3 researchers, each of whom devotes 20 hours to a project, adds up to 60 hours of person time. Similar calculations are made for the persons who are subjects of research.

PERT Program Evaluation and Review Technique. A method for planning a complex activity sometimes used to plan research projects. Originally used by the U.S. Navy to produce the Polaris missile. Sometimes called the "critical path method."

Phenomenology A philosophical doctrine established by E. Husserl in the 19th century that has had considerable influence in the social sciences, particularly sociology. Husserl stressed the rigorously descriptive, but introspective, study of how people perceive and understand the world. *Ethnomethodology is one modern descendent of phenomenology.

Phi [Φ, φ] See *phi coefficient.

Phi Co Short for *phi coefficient.

Phi Coefficient A type of *correlation or measure of *association between two *variables used when both are *categorical and one or both are *dichotomous. Phi is a *symmetric measure. It is based on the *chi-squared statistic (specifically, to get phi you divide chi-squared by the sample size and take the square root of the result). Compare *Cramer's V, which is a better measure when tables have more than four *cells. See also *Pearson's contingency coefficient.

P

For example, to compute the correlation between sex (male/female) and employment status (employed/unemployed) you could use a phi coefficient. You couldn't use it for age and income, because neither of these is dichotomous or categorical.

Pi [Π, π]   (a) A Greek letter used to symbolize a mathematical constant used in important statistics formulas, such as the formula for the normal curve. It is the ratio of the circumference of a circle divided by its diameter, approximately 3.14. (b) Pi is also used to symbolize a *proportion in a *population; the sample proportion is symbolized *P*.

Pie Chart   A circle with areas ("slices") marked to represent the proportion of total units in each category.

The following two pie charts show each continent's percentage of the world's total land area and population.

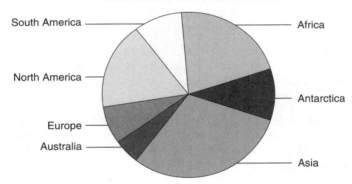

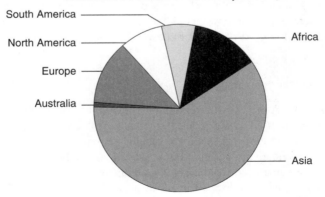

Figure P.3   Pie Chart

Pillias Criterion A *test statistic that is very *robust and not highly linked to *assumptions about the *normality of the *distribution of the data.

Pilot A preliminary test or study to try out procedures and discover problems before the main study begins. This enables researchers to make last-minute corrections and adjustments. It is a research project's "dress rehearsal."

In a pilot, the entire study with all its instruments and procedures is conducted in miniature (e.g., on a small sample). By contrast, a *pretest, definition (b), is used to assess some part of an instrument or procedure.

Placebo (a) In experimental research, a *treatment given to a *control group that is meant to have no effect; it is used in comparison to the treatment (*independent variable) that is being tested. (b) A treatment given to a patient for its psychologically soothing effect rather than for its physiological bene-fits. See *double-blind procedure.

For example, in an experiment on the benefits of a drug, the control group could be given a sugar pill (the placebo) that resembled the drug being tested, while the experimental group would be given the drug.

Placebo Effect Improvement in the condition of sick persons that cannot be attributed to the physiological effect of the treatment that was used, but is due, rather, to their (mistaken) belief that they received an effective treatment (for example, the medicine rather than the sugar pill). Compare *Hawthorne effect.

Planned Comparisons Comparisons between the *means of *experimental and *control groups in *regression analysis or *analysis of variance. So called because they are specified at the outset of the research, before the *data are gathered. Also called "a priori comparisons." Compare *omnibus test, *post hoc comparisons.

The advantage of planned comparisons is that the researcher can move directly to comparisons of interest without first doing an *omnibus test. Such comparisons involve less risk of *Type I error than do post hoc comparisons.

Platykurtic Flatter than a *normal curve. See *kurtosis for an illustration.

Pluralism The belief that research methods often seen as incompatible aren't necessarily so. Quantitative versus qualitative and explanation versus understanding are typical of the dichotomies pluralists reject or think are overemphasized.

Point Biserial Correlation A type of correlation to measure the *association between two *variables, one of which is *dichotomous and the other *con-tinuous. Compare *biserial correlation and *widespread biserial correlation.

Point Estimate An estimate made by computing a *statistic that describes a *sample; this is then used to estimate a *population *parameter. The term

P

estimate, used without specification, almost always means point estimate. Compare *interval estimate, *confidence interval.

For example, you could survey a sample of students in your department to get their opinion about the statistics requirement. Say that on a scale of 1 to 10, the *mean score the sample of students gave the statistics requirement was a rating of 4.0. You could then conclude that 4.0 would be the best estimate of the mean score of the population; that is, it is your best (point) estimate of what the mean score would have been had all the students been polled. An interval estimate might be that the value for the population ranges between 2.0 and 6.0.

Poisson Distribution   A *probability distribution that can be used for a statistical test when the number (N) of cases is very large and the probability (p) is very small.

For example, suppose the murder rate in U.S. cities is .0001, or 100 murders per million residents. Say that Our Town, a city of 500,000 people, had 80 murders last year. A simple calculation shows that Our Town's rate is higher than the typical rate (80/500,000 = .00016, or 160 murders per million). But a more interesting question perhaps is whether 80 murders is higher than what could be expected by chance if ours is a typical town. We could use the Poisson distribution formula to find out. If Our Town is typical, then a total of 80 murders in a year has a probability of about .24. We might conclude that 30 murders more than the average rate could be expected by chance, and that there was nothing atypical about Our Town. However, if the number of murders had been 120 instead of 80, we would probably come to a different conclusion, since the probability of 120 murders in a typical city of 500,000 is only .04.

Poisson Process   A mathematical description of *random events that pertains when events occur randomly in such a way that for each small interval of time, one event or zero events—but not two events—can occur.

Policy Analysis (Research)   The study of social, political, economic, educational, and other policies. The goal is to discover alternative policies and/or problems that require action. Policy analysts are often closely allied with *evaluation researchers. Compare *applied and *basic research.

The following chart shows one way to divide up the nonbasic research domain.

Table P.1   Policy Analysis/Research Compared to Other Types

Type of Research	Typical Question Asked
Policy	What should we do?
Applied	How should we do it?
Evaluation	How well did we do it?

P

Polygon   A line graph drawn by connecting the midpoints of the bars of a *histogram. (From the Greek for "many sided.") Also known as a *frequency polygon; see that entry for an illustration.

Polynomial Equation   An equation in which one or more of the terms is raised to a *power greater than 1. See *linear equation, *polynomial regression analysis.

For example, the regression equation $Y = a + bX + bX^2 + bX^3$ is a polynomial because X is raised to the second and third powers. The highest power of a term gives the equation its "degree" or "order." The example equation is thus a third-order (or degree) polynomial equation.

Polynomial Regression Analysis   *Curvilinear regression analysis, that is, regression analysis for relations that are or are suspected to be nonlinear. So called because the regression equation for a curvilinear (nonlinear) relation is a *polynomial equation. The more turns in the regression line, the higher the power of the terms in the equation must be. Compare *spline regression.

Polytomous Variable   A *categorical variable with more than two categories, literally, with "many divisions"—as in marital status: single, married, divorced, widowed. Also spelled "polychotomous."

Pooled Cross-Sectional Analysis   A type of analysis conducted by combining (pooling) two or more *cross-sectional studies. This pooling is most often done for one of two reasons: to increase overall sample size beyond what is available in any single cross-sectional sample; or to study the effects of the passage of time by comparing samples drawn in different years.

Population   A group of persons (or institutions, events, or other subjects of study) that one wants to describe or about which one wants to generalize. In order to generalize about a population, one often studies a *sample that is meant to be representative of the population. Also called *target population and "universe."

Population Parameter   A characteristic of a population described by a statistic, such as a *mean or a *correlation. Population parameters are usually symbolized by Greek letters; the Roman (English) alphabet is often used for *sample statistics. Consistency is not perfect, however, as Greek letters are sometimes used for both statistics and parameters.

Population Pyramid   A graphic means of describing the sex and age distribution of a population. Like *stem-and-leaf diagrams, they are particularly useful for comparing different distributions. The examples that follow compare two quite different countries. In Mexico the modal age group is 0–4. In the United States, it is 35–39.

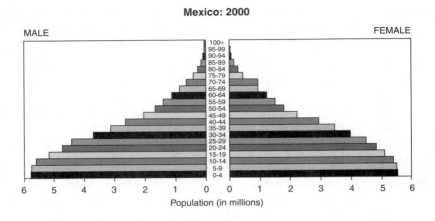

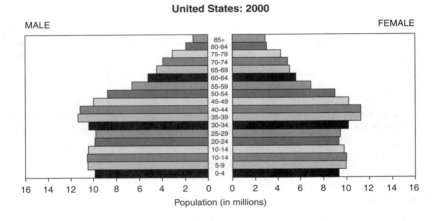

Figure P.4   Population Pyramid

Population Validity   A type of *external validity or representativeness; often used to describe research in which criteria for validity are not met. Compare *ecological validity.

For example, studies in which the samples are composed solely of college sophomores might not be validly generalizable to the population of all adults.

Positive Number   A number greater than zero.

Positive Relation (or Correlation)   Another term for a *direct relationship, that is, one in which increases in one *variable are accompanied by increases in another. A relation is also positive when decreases in one

P

variable are accompanied by decreases in another. Compare *inverse relationship, *negative relation.

Note that positive in this context does not mean beneficial or advantageous. For example, if the crime rate goes up when the poverty rate goes up, the correlation between the two rates is "positive," which means only that the correlation is a positive number, not that crime or poverty or the relation between them is "good."

Positive Skew Said of a graph on which the smaller frequencies are found on the positive (or high or right) end of the *x axis. For example, the distribution of scores on a difficult test would have a positive skew, while the scores on an easy test would be negatively skewed. Also called "right skew." For an illustration, see *skewed distribution.

Positivism A term introduced by Auguste Comte (1798–1857) to refer to the *empirical study of phenomena, especially human phenomena. Comte contrasted the "positive knowledge" gained in this way with the less scientific knowledge obtained by *metaphysics and religion. Compare *scientism.

The term has had several meanings since it was introduced in the early 19th century and repopularized by the Vienna Circle of philosophers in the 1930s (see *logical positivism), and it has always been the object of debate. Most commonly and loosely today, positivism refers to a belief, held by some people, that one can study scientifically and/or quantitatively things that other people believe cannot or should not be studied in this way, such as religion, emotions, ideas, art, and morality.

Posterior Probability In *Bayesian inference, an investigator's opinion, expressed as a probability, after research has been done assessing the investigator's *prior probability opinions.

Post Hoc Comparison A test of the *statistical significance of differences between group *means calculated after ("post") having done an *analysis of variance (ANOVA) or a *regression analysis (RA) that shows an overall difference. See *planned comparison, *omnibus test, *fishing expedition. Also called "post hoc test," "a posteriori comparison," and "follow-up test or comparison."

The *F ratio of the ANOVA tells you that some sort of statistically significant differences exist somewhere among the groups being studied. Subsequent, post hoc analyses are meant to specify what kind and where. Since these comparisons are not part of the original study design (otherwise they would be *planned comparisons), some researchers consider them of doubtful validity. Such comparisons are easiest to justify for *exploratory research; they should be used sparingly to test *hypotheses.

Post hoc comparison is a generic term for several kinds of analyses. *Tukey's HSD and the *Scheffé method are two of the best known. Others include the *Newman-Keuls test and the *Duncan Multiple-Range test. Of the four, Duncan's is the least *conservative measure; Scheffé's is the most conservative. Tukey's is more often associated with ANOVA, Scheffé's with RA. In any case, the general procedure is the same: After you get a significant overall or omnibus test result, you look for differences among groups to explain it. Because the comparisons are multiple and carry with them increased risk of *Type I error, the *critical value of the test statistic is raised to compensate and thus to reduce that risk.

Post Hoc Theorizing   "After this" theorizing, that is, existing data are used to "test" theories that were originally constructed to describe those very same data. While this would be bad practice in experimental research, more latitude may be justifiable in nonexperimental work, especially exploratory studies. But in all cases good practice is to test a theory with a new set of data, not the data used to develop the theory.

Postmodernism   A broad term covering a multitude of loosely related beliefs that have in common a distrust of "modern science," and more broadly, "modern" ways of thinking that stress individual rationality. Its adherents tend to be especially critical of *positivism. Perhaps the most common trait of postmodernists is *relativism.

Posttest   A test given or measurement taken after an experimental *treatment. Compare *pretest, with which the results of a posttest are usually compared.

Postulate   A conjecture or other statement considered to be an essential starting point in a chain of logical reasoning. While there can be subtle distinctions, it is generally accurate to say that *assumption, *axiom, and postulate refer to the same basic concept.

Power   The number of times a number is multiplied times itself, usually written as an *exponent. See *polynomial equation.
   For example, 8^2 is 8 to the second power and means $8 \times 8 = 64$; 8^3 is 8 to the third power and means $8 \times 8 \times 8 = 512$.

Power of a Test   Broadly, the ability of a technique, such as a statistical test, to detect relationships. Specifically, the probability of rejecting a *null hypothesis when it is false—and therefore should be rejected. The power of a test is calculated by subtracting the probability of a *Type II error from 1.0. The maximum total power a test can have is 1.0; the minimum is zero; .8 is often considered an acceptable level for a particular test in a particular study. Also called *statistical power.
   Power analyses are sometimes conducted *before* data are gathered in order to determine how large a sample needs to be to avoid Type II error. They are also conducted *after* the data have been gathered; the data and the sample size are then used to estimate the probability that a Type II error has been committed.

Power Transformation   *Transformation of a variable by taking a power of its values, as by squaring or cubing it.

PR   An abbreviation for *probability.

Practical Significance   Said of a research finding that one can put to use, that can change practice. Usually contrasted with (mere) *statistical significance; closely related to *substantive significance. "Clinical significance" refers to practical significance in a clinical setting, for example, an effect that is large enough to be a reasonable treatment option. There are no statistical tests for practical significance. Whether something is practically significant is a matter of judgment.

Often researchers talk of a low level of practical significance to describe a finding that is too large to be likely due to chance alone (i.e., one that is statistically significant), but not large enough to be of use. Sometimes, however, the practical significance of a finding can be great even when it is not large. For example, the statistical *association between amount of exercise and longevity is quite modest. But it is practically significant for two reasons: (1) longevity is important (literally a matter of life or death), and (2) unlike many other variables that affect longevity, we have considerable control over the amount we exercise.

Practice Effects   Influences on the outcomes of a study that occur when subjects are tested more than once. Subjects may get adept at the task as a result of such practice—or bored with it and consequently less proficient.

Pragmatism   A philosophical movement that grew up in the United States in the late 19th century that had a good deal of influence on how social scientists viewed their work. Skeptical of broad truths (see *fallibilism), pragmatists held that knowledge is defined by usefulness; the ultimate test of a proposition is whether it "works," particularly in helping individuals solve practical problems.

PRE   Proportional reduction of error. A type of *measure of association that calculates how much you can reduce your error in the prediction of a *variable by knowing the value of another variable. How much better can you estimate Y if you know X than if you do not? The answer is a PRE statistic. *Lambda is one example of a PRE measure; *Pearson's r is not, but the r^2 is.

Precision   The degree to which measurements are replicable or consistent, usually determined by how much they are subject to *random error. Compare *reliability. See *specificity, *sensitivity. Contrast *validity, which means accuracy.

Predetermined Variable   A variable in a *path analysis the cause or causes of which are not specified. Such variables are usually connected to other predetermined variables with curved, two-headed path lines. Also called *exogenous variables. See *path diagram for an illustration.

Prediction   The term prediction has two common meanings in research: (a) using data to make a statement about the future, as in *forecasting; (b) the more common use of the term prediction, and the one that sometimes confuses novice researchers, refers to using data to "predict" outcomes that have already occurred, as in most *regression analyses. For example, one might use data about members of the entering class of 1999 to "predict" their likelihood of graduation in 2003. Obviously, in such cases one is studying statistical associations among variables in the past, not "predicting" in the ordinary sense of the term. One could, however, use the equation developed by studying past data about the class of 1999 to then predict graduation rates of future classes.

Prediction Equation   Another term for a *regression equation, that is, an equation that predicts the value of one variable on the basis of knowing the value of one or more other variables. A prediction equation is a regression equation that does not include an *error term. See *PRE.

Predictive Research   Said of an investigation whose goal is to forecast (predict, but not explain) the values of one variable by using the values of one or more other variables. Usually contrasted with *explanatory research, in which the goal is to understand the *causes behind relations. In other terms, the goal in predictive research is to estimate a future value of a dependent variable; in explanatory research it is to estimate the values of *independent variables.

  For many authors writing about research methods, the distinction between explanatory and predictive research usually amounts to a dichotomy between *experimental research (which can be explanatory) and other kinds, especially *correlational research (which cannot). The debate over whether only experimentalists can make valid causal generalizations, and thus do explanatory research, is sometimes quite heated and more than a little self-interested.

Predictive Validity   The extent to which a test, scale, or other measurement predicts subsequent performance or behavior. Also called *criterion-related validity.

Predictor Variable   Loosely, another name for *independent variable or *cause. The term is often used when discussing *nonexperimental research designs such as *correlational studies. Also called *explanatory variable. Compare *criterion variable.

PRESS Statistic   Predicted residual sum of squares. A technique used to estimate how widely a *multiple regression model can be generalized and to compare models. Values range from zero to 1.0, with those closer to 1.0 indicating a higher level of *generalizability.

P

Pretest (a) A test given or measurement taken before an experimental *treatment begins; more generally, a "before" measure in a before-and-after study. By contrasting the results of the pretest with those of the *posttest, researchers gain evidence about the effects of the treatment. (b) A trial run used to assess some part of an *instrument or procedure. Compare *pilot.

Pretest Sensitizing Unintentionally influencing the results of the *posttest by giving a pretest. Compare *practice effects, *order effects.

For example, the pretest could influence the posttest results if subjects have been alerted to what the researcher is interested in, or if they learn skills on the pretest that they can later use on the posttest.

Prima Facie True on first view; true unless demonstrated otherwise by subsequent evidence.

Primary Analysis Original analysis of the data in a research study. Compare *secondary analysis (reanalysis of data gathered by others) and *meta-analysis (analysis of analyses).

Primary Source An original source of *data; one that puts as few intermediaries as possible between the production and the study of data. Compare *secondary source.

For example, the primary sources for the study of the use of metaphor in Shakespeare would be works written by Shakespeare; secondary sources would be books about Shakespeare. "Primary" is often more a matter of degree than kind; for instance, Shakespeare's original manuscripts would be more primary than a 20th-century edition of his works.

Principal Components Analysis (PCA) Methods for undertaking a *linear transformation of a large set of correlated variables into smaller uncorrelated groups of variables. This makes analysis easier by grouping data into more manageable units and eliminating problems of *multicollinearity. Principal components analysis is similar in aim to *factor analysis, but it is an independent technique; advocates of the two methods stress the differences rather than the similarities, and debates between them are often highly contentious. However, the outcomes produced by the two methods are usually quite similar, and PCA is often used as a first step in factor analysis. See *discriminant analysis, *canonical correlations, *cluster analysis.

Prior Probability In *Bayesian inference, the opinion of an investigator who states his or her opinion as a probability before (prior to) a research study. On the basis of the study, she or he revises the opinion in light of the data collected in order to arrive at a new *posterior probability. Also called "priors."

Priors (a) *Prior probabilities or prior distributions in *Bayesian inference. (b) *Prior variables in a causal model.

P

Prior Variable   Another term for *exogenous variable.

Prisoners' Dilemma   In *game theory, a classic illustration of competition when information is imperfect. The situation involves two prisoners accused of committing a crime together. They are questioned separately. If both refuse to confess, they will do better than if both confess. If one confesses and the other does not, the one who confesses goes free while his partner goes to jail. A remarkable number of findings have been built upon the mathematical and logical analyses of this and similar simulations of competition.

Probabilistic   (a) Said of research that focuses on *probabilities rather than, for example, universal *laws or truths. (b) Said of a causal relationship in which change in one *variable increases the *probability of change in another variable, but does not invariably produce the change. Compare *necessary condition, *sufficient condition.

For example, marriage and sexual intercourse increase the probability of parenthood, but they do not invariably cause it.

Probability   Probability, like "cause," is a highly controversial concept; there are different—and opposed—camps of statisticians studying probability, most importantly the *frequentists and the *Bayesians. The definitions that follow generally cover both. (a) The likelihood that a particular event or relationship will occur; the *proportion of tries that are *successes. More formally, out of all possible outcomes, the proportionate expectation of a given outcome. Values for statistical probability range from 1.0 (always) to 0 (never). Probability statistics cannot be negative numbers, because something cannot be less than totally unlikely. See *p value, *likelihood ratio, *Bayesian inference. (b) The field of mathematics devoted to the study of probability as defined in definition (a).

For example, the probability that a person drawing a card at random from a deck of 52 playing cards will select a red card is 26/52, or .5. The probability of drawing a heart is 13/52, or .25; for an ace it is 4/52, or .077.

Probability, Conditional   Said of situations in which the probability of one event (A) depends on (is conditioned by) another event (B). Symbolized $p(A|B)$. Contrast *independence.

For example, the probability of contracting the AIDS virus is conditional upon contact with an infected person—$p(AIDS|CONTACT)$. The greater the number of such contacts, the greater the probability of getting the disease.

Probability, Empirical   An actual count of the number of events of a particular type divided by the total number of possible events. Usually contrasted with *theoretical probability.

For example, if we had a perfectly balanced coin, the theoretical probability of tossing the coin and getting tails is .50, that is, 1 out of 2 (1/2). But

P

if we actually flipped the coin, say, 200 times, we might get 96 tails and 104 heads. The empirical probability—describing what actually happened— would be .48 for tails, that is, 96 out of 200 (96/200 = .48). As we increase the number of trials (coin flips), the empirical probability tends to get closer to the theoretical probability.

Probability, Joint   The probability of two or more events occurring together.
For example, the probability of drawing a spade from an ordinary deck of 52 playing cards is 13 out of 52, or .25. The probability of drawing a jack is 4 out of 52, or .0769. The joint probability of drawing a card that is both a jack and a spade is calculated by multiplying the probability of a spade (.25) times the probability of a jack (.0769). The product, .01923 (.25 × .0769), is the joint probability of the jack of spades.

Probability, Subjective   A guess or feeling about some probability that is not based on any precise computation. But it may be a reasonable, if not computational, assessment by a knowledgeable person. See *Bayesian inference.
For example, an experienced member of a parole board might say, "I really believe that, on the basis of his attitude and his record in the work release program, the prisoner is a good risk; we should give him a chance." Or the loan committee at a bank might conclude, "Given that several competing businesses are nearby and that previous establishments at this location have gone out of business, we feel that the probability of failure is high and, therefore, we recommend against the loan."

Probability, Theoretical   The expected or predicted likelihood that a particular event will occur. See *probability, empirical.

Probability Density Function   A distribution of measures of probability for a continuous variable. The measures are called "densities," and the "function" is a formula for drawing a *histogram depicting the densities. See *probability distribution, definition (b), and *frequency curve.

Probability Distribution   (a) All of the outcomes in a distribution of research results and each of their probabilities (*empirical probability distribution). (b) The number of times we would expect to get a particular number of *successes in a large number of trials (theoretical probability distribution). The most important of the theoretical probability distributions are the *binomial, *normal, *Student's t, *chi-squared, and *F distributions. Theoretical distributions are compared with observed, empirical distributions in order to judge the probability that the latter could have occurred by chance alone.
For a *continuous variable, an empirical probability distribution is based on intervals. For example, a probability distribution for income could be the percentage of households whose incomes fell in each of the following

P

intervals: less than $25,000 per year; $25,001 to $30,000; $30,001 to $35,000; and so on. For *discrete variables, the probabilities are *relative frequencies. See *binomial distribution for an example.

Probability Level   The *p value below which the *null hypothesis is rejected; this value or level is the chance of making a *Type I (or *alpha) error.

Probability Paper   Graph paper calibrated so that the values of a *cumulative frequency distribution fall on a straight line if the data being graphed are normally distributed. Less used today than computerized graphics equivalents. See *normal probability plot, *P-P plot.

Probability Plots   Graphic techniques for comparing probability distributions, usually to determine whether a data set is normally distributed. See *normal probability plot for graphics; also *probability paper, *P-P plot, *Q-Q plot.

Probability Sample   A *sample in which each case that could be chosen has a known probability of being included in the sample. Often a *random sample, which is an equal probability sample. At some point, *random selection is part of the process of every probability sample.

Probability Statistics   Techniques for calculating the likelihood (*probability, usually expressed as a p value) that particular events will occur.

Probability statistics are the bridge from *descriptive statistics about *samples to *inferential statistics about *populations. After calculating a descriptive statistic for a sample, one then calculates the probability that this statistic could have been obtained by chance alone. If that probability is low, then one can reasonably infer that what is true of the sample is probably true of the population as well. See *sampling distribution, *statistical inference.

Probability Theory   A branch of mathematics dealing with how to estimate the chances that events will occur. Compare *decision theory, *risk.

In a way, probability theory, especially the *frequentist variety, is *inferential statistics stood on its head. In probability theory, we make inferences about samples or individual events on the basis of knowledge about populations, but in inferential statistics, we use samples to generalize about populations. For example, if we know that a particular type of surgery is successful over 90% of the time, we might generalize to individual patients to say that the probability is over .90 that it will be successful for them.

Probability Value   The likelihood that a statistical result would have been obtained by chance (or *sampling error) alone. This actual probability value (p value) is compared by a researcher with an *alpha level (sometimes called *probability level) to determine whether the result has *statistical significance. If the p value is smaller than the alpha level, the result is judged to be statistically significant. For example, if the alpha is .05 and the calculated p is .022, the result is declared statistically significant, $p < .05$.

Probit Regression Analysis   A technique used in *regression analysis when the *dependent variable (DV) is a *dummy (or *dichotomous) variable. In the probit model, *independent variables (IV) influence an unobserved continuous variable, which in turn determines the probability that an observed dichotomous event occurs. Probit is short for "probability unit." See *logit. Compare *logistic regression.

To conduct a probit regression analysis, one first needs to *transform the categorical DV. The probit and *logit provide roughly equivalent (though mathematically quite different) ways of doing this. For regressions with dummy DVs, the logit and probit approaches almost always give very similar results. In brief, in probit regression one transforms the probabilities of the DV (e.g., of graduating) into z-scores, using the *cumulative standard normal distribution (CSND) to do this. Details aside, the questions you answer with probit regression are variants (in probability units) of the questions you answer with any regression analysis. Those questions in probit regression language are: (1) What is the change in the probability of the DV being 1 (e.g., yes, graduate) for a one-unit change in an IV, while controlling for the effects of the other IVs? (2) What is the cumulative effect of all the IVs together on the probability of the DV equaling 1 (yes)?

Procrustean Transformation or Rotation   A technique used in *factor analysis to transform an empirical matrix so that it resembles a predetermined target matrix, such as a theoretically expected pattern. Often used in conjunction with *Promax rotation.

Product   The result of multiplying. For example, in the equation $9 \times 6 = 54$, 54 is the product.

Production Function   An equation expressing the relationship between the output of a good and the inputs needed to make it. Usually a form of *regression equation. See *productivity.

Productivity   Output per unit of input. The concept, originally developed in economics, has wide applicability in other fields.

Product-Moment Correlation   See *Pearson's correlation coefficient (r).

Product Variable   A variable obtained by multiplying the values of two other variables. Product variables are most often used to study *interaction effects. See *product vector.

Product Vector   The result of multiplying two *vectors of scores on variables. It is a step on the way to calculating the *covariance of two variables. *Cross-product vector is another term for the same operation.

For example, in the following table Vector 1 is multiplied by Vector 2 to get Vector 3, which is the product vector.

Table P.2   Product Vector

Case	Vector 1	Vector 2	Vector 3
1	2	4	8
2	3	2	6
3	5	6	30
4	7	8	56

Program   A set of instructions, written in a *programming language, that tells a computer how to handle *data according to certain rules. Unless it is programmed, a computer cannot perform operations. Also called *software.

Program Evaluation   See *evaluation research.

Programming   Writing a set of instructions to solve a problem with a computer.

Programming Language   A *software program used to write other programs. Among many examples, some of the best known include FORTRAN, BASIC, and C.

Projective Tests   Psychological tests in which subjects are asked to describe what they see in pictures that can be interpreted in many ways. The idea is that subjects' interpretations will "project" their inner states. The Rorschach ("inkblot") test is a well-known example.

Promax   A method of *oblique rotation of the axes in a *factor analysis.

Proportion   A number, ranging between 0 and 1.0, calculated by dividing the number of subjects having a certain characteristic by the total number of subjects. To get a percentage, you multiply the proportion by 100. See *relative frequency, *probability.

For example, if the graduating class had a total of 948 students and 237 of those were business majors, the proportion of business majors would be .25 (237/948). Twenty-five percent (.25 × 100 = 25%) are business majors.

Proportional Reduction of Error (PRE)   A type of measure of *association that indicates how much you can reduce the error in the prediction of one variable by knowing the value of another variable. How much better can you estimate variable Y if you know variable X? The answer is the proportional reduction of error. *Gamma, R^2, omega squared, eta squared, and *lambda are examples of PRE measures of association.

Proportional Stratified Random Sample   A *stratified random sample in which the proportion of subjects in each category (stratum) is the same as in the *population. Compare *quota sample.

Proportion of Variance Index   Another term for a *proportional reduction of error (PRE) measure.

Proposition   A formal statement about the relationships among abstract concepts. Propositions usually occur early in an article or other research report. Subsequent parts of the report generally try to maintain or demonstrate the proposition. A *theory is a set of related propositions.

     For example, "cognitive sophistication fosters social tolerance" is a proposition.

Prospective Study   A *longitudinal study in which subjects are followed to see what will happen to them. Also called *panel study. The opposite would be a *retrospective study. The term is more commonly used in epidemiology than in the social sciences.

Protocol   The plan for carrying out research, especially for administering *experimental treatments. Sometimes a distinction is made between the *design, which is the overall plan, and the protocol, which is the document detailing the step-by-step procedures. Detailed protocols are especially important for reducing *random error and *bias when researchers hire others to collect their data, and in multisite research where treatments are administered by many different researchers.

Proxy Variable (or Measure)   An indirect measure of the variable a researcher wants to study; it is used when the object of inquiry is difficult to measure or observe directly. See *indicator, *operation. Also called "surrogate variable."

     For example, in a classic study, Durkheim wanted to study the historical evolution of moral values in European societies. Since there was little directly available objective evidence about moral values, he studied the history of law. Laws, he believed, reflect moral values, and laws are easily accessible to the researcher.

Pseudo R^2   A measure used in *logistic and *probit regression to represent the proportion of error variance controlled by the model. By contrast, the "true" R^2 of *OLS regression represents the proportion of the variance in the dependent variable predicted by the independent variables. Since there can be no variance from the mean with categorical variables (because there is no meaningful mean), the R^2 estimate is spurious or "pseudo." There is more than one way to estimate the pseudo R^2, and the different methods can produce quite different values, so much so that some statisticians argue against using any of them.

P

Pseudo-Random Numbers   Numbers that approximate a set of true random numbers, usually as produced by a mathematical formula as generated by a computer. Random numbers have wide uses in statistics—in *simulations, for example—and the demand for tables of random numbers is high.

Psychometric Research   Research on how psychological *variables are *operationalized for purposes of measurement, particularly measurement of individual differences among people.

Public Use Microdata Samples (PUMS)   Large samples of data from the U.S. Census Bureau's decennial censuses. They are available to any researcher and are designed for doing *secondary analyses with a personal (micro) computer. Names, addresses, and other identifying data are deleted before the PUMS are made available to researchers.

Publication Bias   In *meta-analysis, the bias that arises when studies showing statistically significant results are more likely to get published and thus be more available for meta-analytic studies. See *file drawer problem.

PUMS   *Public Use Microdata Samples.

Pure Research   Another term for *basic research; the opposite is usually *applied research.

Purposive Sample   A *sample composed of subjects selected deliberately (on purpose) by researchers, usually because they think certain characteristics are typical or *representative of the *population. Compare *quota sample.

   This is generally an unwise procedure; it assumes that the researcher knows in advance what the relevant characteristics are; it runs the risk (because it is not *random) of introducing unknown *bias. Inferences about a population cannot legitimately be made using a purposive sample. On the other hand, purposive sampling is often the only way to try to increase representativeness in *field research, and it can be an improvement over simple *convenience sampling. A frequent compromise between *random sampling and purposive sampling is *stratified random sampling.

PUS   Public Use Samples. See *Public Use Microdata Samples.

p **Value**   Short for *probability value, that is, the probability that a statistic could occur by *sampling error—if the null hypothesis is true.

   For example, if a correlation between two variables in a sample of 100 cases were reported as $r = .43$, $p = .018$, this would mean the following: If there were no correlation in the population from which the sample was drawn (the null hypothesis), the probability of obtaining a correlation this size (.43) in a sample this size (100) would be .018 Researchers often round off a *p* value such as .018 to < .05, but the actual *p* value is more informative.

Pygmalion Effect Changes in subjects' behaviors brought about by researchers' expectations. See *self-fulfilling prophecy, *demand characteristics. Also called "Rosenthal Effect."

The term originally comes from Greek mythology; it was popularized in a play by G. B. Shaw; it is perhaps best known to researchers in the form of a controversial study (by Rosenthal and Jacobson) in which teachers were told to expect some of their students' intelligence test scores to increase. They increased, apparently because of teachers' expectations.

Q (a) Abbreviation for *quartile. Q1 is the first quartile, Q2 the second, and so on. (b) Symbol for the *studentized range statistic. (c) A measure of *goodness-of-fit for an *overidentified *path model. It ranges from 0 to 1, with 1.0 indicating a perfect fit. (d) *Yule's Q. (e) *Cochran's Q Test.

Q Methodology A way of ordering or sorting the parts of objects of study. The most recognizable feature of Q methodology is that "subjects" sort cards (a Q sort) into a number of piles that represent points on a continuum. The sorters are most often not the subjects of the research; they are *respondents or even *informants. They sort statements into categories based on their personal understanding of the concepts being investigated.

For example, one might ask known conservatives and liberals to sort a group of cards on which are written the names of government policies. They would rank the policies on a scale ranging from most to least favorable. The researcher could use their rankings to test a theory of the components of liberalism and conservatism by seeing how the two groups of sorters differed.

Q-Q Plot Quantile-Quantile plot. A method for comparing a sample distribution with a theoretical distribution, such as a *normal curve, by plotting them against one another. This is usually done with *percentiles as the quantiles. If the sample has been drawn from a normal distribution, the plotted values of the sample will approximate a straight line. See *normal probability plots for an illustration.

Q Sort See *Q methodology.

Quadratic Relation Said of a relation in a *regression analysis when one of the *independent variables has been raised to a power of 2 (X^2) and doing so produces a significant *regression coefficient. A quadratic relation indicates that there has been one departure from *linearity, that is, one turn or change of direction in the regression line. See *curvilinear relation, *polynomial regression analysis, *spline regression.

Qualitative   (a) When referring to *variables, qualitative is another term for *categorical or *nominal. (b) When speaking of kinds of research, qualitative refers to studies of subjects that are hard to quantify, such as art history. The term qualitative research tends to be a residual category for almost any kind of nonquantitative research. See *participant observation, *ethnographic research, *focus group.

The qualitative-quantitative distinction is often overdrawn. It is difficult to avoid *quantitative elements in the most qualitative subject matter. For example, "The painter entered his 'blue period' in the 1890s." And qualitative components are crucial to most good quantitative research, which begins with *theories, *concepts, and *constructs.

Qualitative Designs   Said of research designs commonly used to study qualitative data. The distinction between qualitative and quantitative design is hard to maintain. Virtually every major research design can be employed to gather either qualitative or quantitative data. For example, surveys are usually thought of as a quantitative design or method, but that is not necessarily the case. Surveys might ask respondents to answer questions on a Likert scale, which are then summed into a quantitative index. But respondents could just as easily be asked open-ended questions; their answers would become texts that could be studied with *grounded theory, which is a qualitative method of analysis.

Quantile   Any of several ways of dividing the total number of rank-ordered cases or observations in a study into equally sized groups—into groups having the same quantity.

*Quartiles, *quintiles, *deciles, and *percentiles are examples of quantiles. Quantiles indicate the proportion of scores located below (and above) a given value. For example, if you scored at the 80th percentile, your score would be above that of 80% of those who were tested.

Quantitative   Said of variables or research that can be handled numerically. Usually contrasted (too sharply) with *qualitative variables and research. Many research designs lend themselves well to collecting both quantitative and qualitative data, and many variables can be handled either qualitatively or quantitatively. For example, *naturalistic observations can give rise to either or both kinds of data. Interactions can be counted and timed with a stopwatch or they can be interpreted more holistically.

Quartile Deviation   Half of the range covered by the middle half of the scores in a *distribution; half of the distance between the first and third *quartiles. The advantage of the quartile deviation, in comparison to other measures of *dispersion, is that it is less influenced by *outliers (extreme values), certainly less so than the ordinary *range. Also called "semi-interquartile range."

Quartiles Divisions of the total rank-ordered cases or observations in a study into four groups of equal size. Technically, the three points that divide a series of ordered scores into four groups. Loosely, the groups themselves. The first quartile is located at the 25th percentile, the second at the 50th, and the 3rd at the 75th. Compare *quantile, *quintile, *decile.

Quasi-Experiment A type of research design for conducting studies in *field or real-life situations where the researcher may be able to manipulate some *independent variables but cannot randomly assign subjects to *control and *experimental groups. See *field experiment, *interrupted time-series design.

For example, you cannot cut off some individuals' unemployment benefits to see how well they could get along without them or to see whether an alternative job training program would be more *effective. But you could try to find volunteers for the new job training program. You could compare the results for the volunteer group (*experimental group) with those of people in the regular program (*control group). The study is quasi-experimental because you were unable to assign subjects randomly to *treatment and control groups.

Quasi-Independent Variable A variable treated statistically as independent, but which cannot be manipulated by the researcher.

Quasi-Random Sample Another term for *systematic sample.

Questionnaire A group of written questions to which subjects respond. Some researchers restrict the use of the term "questionnaire" to written responses.

Quetelet Index The *body mass index (BMI), named after its 19th-century Belgian creator Adolphe Quetelet (1796–1874).

Queuing Theory The mathematical study of waiting lines or queues. It is used most widely in economics and business and can be applied to any waiting line from customers at checkout counters to aircraft on runways.

Queue Processing See *first in-first out (FIFO).

Quintiles Divisions of the total rank-ordered cases or observations in a study into five groups of equal size. Technically, the four points that divide the observations into five groups. Loosely, the five groups themselves. Compare *quantile, *quartile, *decile.

Quota Sample A *stratified *non*random sample, that is, a sample selected by dividing a *population into categories and selecting a certain number

Q

(a quota) of respondents from each category. Individual cases within each category are not selected randomly; they are usually chosen on the basis of convenience. Compare *accidental sampling, *purposive sample, *random sampling, *stratified random sampling, *proportional stratified random sampling.

For example, interviewers might be given the following assignment: "Go out and interview 20 men and 20 women, with half of each 50+ years old." Despite its superficial resemblance to *stratified random sampling, quota sampling is not a reliable method to use for making inferences about a population.

Quotient   The number that results when one number is divided into another; the answer to a division problem.

r   Symbol for *Pearson's correlation, which is a *bivariate correlation (between two variables).

R   (a) Symbol for a *multiple correlation, that is, among more than two variables. (b) Abbreviation for *range.

R   A statistical software package available to download at no cost from the manufacturer. It is modeled after *S-PLUS.

r²   Symbol for a *coefficient of determination between two variables. It tells you how much of the *variability of the *dependent variable is explained by (or accounted for, associated with, or predicted by) the *independent variable. Sometimes written "r-squared."

For example, if the r^2 between students' SAT scores and their college grades were .30, that would mean that 30% of the variability in students' college grades could be predicted by variability in their SAT scores—and that 70% could not.

R²   Symbol for a *coefficient of multiple determination between a *dependent variable and two or more *independent variables. It is a commonly used measure of the *goodness-of-fit of a *linear model. Sometimes written "R-squared."

For example, if the R^2 between average individual income (the dependent variable) and father's income, education level, and IQ were .40, that would mean that the effects of father's income, educational level, and IQ together explained (or predicted) 40% of the variance in individuals' average incomes—and that they did not explain 60% of the variance.

***r*₍bis₎**   *Biserial correlation

***r*₍pb₎**   *Point-biserial correlation.

r_s Symbol for the *Spearman correlation coefficient (rho).

r_{tet} *Tetrachoric correlation.

R_c Symbol for the *canonical correlation.

Radical   The *root of a number as shown by the radical sign $\sqrt{}$. A number (the "index") to the left of the sign shows the type of root. For example, $\sqrt[3]{}$ means the third (cube) root. If there is no number, the root is a square root.

Radix   In a *life table, the starting number (100,000 in the example at the life table entry). The number perishing at different ages is subtracted from the radix to obtain the number surviving.

R and D   *Research and development.

Random   Said of events that are unpredictable because their occurrence is unrelated to their characteristics; they are governed by chance. The opposite of random is determined.

The chief importance of randomness in research is that by using it to select or assign subjects, researchers increase the probability that their conclusions will be *valid. *Random assignment increases *internal validity. Random sampling increases *external validity. See *probability sample, *pseudo-random numbers.

Random Assignment   Putting subjects into *experimental and *control groups in such a way that each individual in each group is assigned entirely by chance. Otherwise put, each subject has an equal probability of being placed in each group. Using random assignment reduces the likelihood of *bias. Also called "random allocation."

Random Coefficients Models   See *hierarchical linear models.

Random-Effects Model   An *experimental design in which the *levels of the *factors are random, in the sense that they are drawn at random from a population of levels rather than fixed by an investigator. Also called "variance components model" and "Model II ANOVA design." Random-effects models are much less widely used than *fixed-effects models in which researchers set the treatment levels. Compare *mixed-effects model, *random variable.

The random-effects model is used when there is a large number of categories or levels of a factor. For example, say researchers in a survey organization wanted to see whether different kinds of telephone interviewers get different response rates. Because there are potentially a very large number of categories (differences in accent, quality of voice, etc.), perhaps as many as there are individual telephone interviewers, a sample is chosen randomly from the population of interviewers, which is also, in this case, a population of levels. On the other hand, if the survey organization were only

R

interested in, say, the difference in response rate between male and female interviewers, they would used a *fixed-effects model.

Random Error   Another term for *random variation. Also called "unrelia-bility" and *disturbance. See *reliability. "Error" without qualification usually means random error. Random errors are often assumed to have a *normal distribution. Random error cannot be eliminated, but it can be esti-mated. The opposite of random error is *systematic error, or *bias, which is usually more difficult to estimate.

Random Factor   A *factor in an *ANOVA in which the levels of the vari-able are points along a continuum, such as 5, 10, 15, and 20 minutes. See *random-effects model.

Randomized-Blocks Design   A *research design in which subjects are matched on a *variable the researcher wishes to control. The subjects are put into groups (blocks) of the same size as the number of *treatments. The members of each block are assigned randomly to different treatment groups. Compare *Latin square, *repeated-measures ANOVA.

For example, say we are doing a study of the effectiveness of four meth-ods of teaching statistics. We use 80 subjects and plan to divide them into 4 treatment groups of 20 students each. Using a randomized-blocks design, we give the subjects a test of their knowledge of statistics. The four who score highest on the test are the first block, the next highest four are the second block, and so on to the 20th block. The four members of each block are randomly assigned, one to each of the four treatment groups. We use the blocks to equalize the variance within each treatment group by making sure that each group has subjects with a similar prior knowledge of statistics.

Randomized Clinical Trial (RCT)   Experimental methods applied to stud-ies of the effectiveness of treatments, originally drug treatments in medical research. The first published results from an RCT appeared in a 1948 arti-cle; it reported the effects of streptomycin on tuberculosis. *Random assignment is used to form treatment and control groups, and *double-blind procedures are applied when possible. Randomized *clinical* and *control* tri-als are sometimes used interchangeably.

Randomized Control Trial (RCT)   Experimental methods applied to the study of the effectiveness of treatments, especially treatments not adminis-tered in a clinical or laboratory setting, such as a *field experiment of an educational intervention. Subjects are placed in *control and *treatment groups by *random assignment. Originally developed in medical research, the methods of the RCT are increasingly demanded in other fields, such as education, and in program *evaluation research more generally.

Randomized Field Trial   Another term for *randomized control trial.

Random Numbers   A sequence of numbers in which the occurrence of any number in the sequence is no guide to the numbers that come next. In the long run, all numbers will appear equally often. Tables of random numbers, found in most statistics texts, are frequently used to select random samples or assign subjects randomly. Random numbers generated by a computer are *pseudo-random, because they are produced according to a formula.

Random Number Generator   A computer *program designed to produce a series of random numbers. These programs usually produce numbers in cycles, and thus generate numbers that are only *pseudo-random.

Random Process   A means by which random numbers or events are generated. Rolling a pair of fair dice is an example.

Random Sampling   Selecting a group of subjects (a *sample) for study from a larger group (*population) so that each individual (or other *unit of analysis) is chosen entirely by chance. When used without qualification (such as *stratified random sampling), random sampling means "simple random sampling." Also sometimes called "equal probability sample," since every member of the population has an equal *probability of being included in the sample. A random sample is not the same thing as a haphazard or *accidental sample. Using random sampling reduces the likelihood of *bias. Compare *probability sample, *cluster sample, *quota sample, *stratified random sample.

Random Selection   Another term for *random sampling. "Selection" is more often used in experimental research; "sampling" is the more common term in survey research, but the underlying principle is the same. See *random assignment, *probability sample.

Random Variable   A variable that varies in ways the researcher does not control; a variable whose values are randomly determined. "Random" refers to the way the events, values, or subjects are chosen or occur, not to the variable itself. Men and women are not random, but sex could be a random variable in a research study; the sex of subjects included in the study could be left to chance and not controlled by the researcher. Also called *stochastic variable.

Random Variation   Differences in a variable that are due to chance rather than to one of the other variables being studied. Random variations tend to cancel one another out in the long run. Also called *random error.

For example, say you take two random samples of 100 workers from a local factory. You find that in the first sample 52 are women and 48 men. In the second sample 49 are women and 51 men. The differences between the sex compositions of the two samples would be due to random variation. See *sampling error.

R

Random Walk (a) A series of steps in which the direction and length of each step is uninfluenced by the previous steps. (b) The random walk hypothesis states that, because stock prices move randomly, in the long run an investor will do better choosing stocks at random (taking a random walk through the market) than by any other method.

Range A measure of *variability, of the spread or the *dispersion of values in a series of values. To get the range of a set of scores, you subtract the lowest value or score from the highest.

For example, if the highest score on the political science final were 98 and the lowest 58, the range would be 40 (98 − 58 = 40).

Range Effects *Floor and/or *ceiling effects.

Rank Correlation See *Spearman's rho and *Kendall's tau, which are the two most frequently used rank correlations.

Rank Data Data measured on an *ordinal (or *rank order) scale.

Rank-Difference Correlation Another term for *Spearman's rho. Compare *Kendall's tau.

Rank Order Scale Another term for an *ordinal scale, that is, one which gives the relative position of a score in a series of scores, such as first in the league, 18th in the graduating class of 400, and so on.

Ranks Numbers assigned to a group of scores arranged in order, such as 1 for the highest score on a test, 2 for the second highest, and so on. Many *nonparametric tests are designed for use with ranks.

Rasch Modeling An early version of *item response theory (IRT) still much in use today. Named after its founder G. Rasch. Debates between proponents of Rasch and IRT models can be quite heated. Both involve methods for assessing the difficulty of test items and the ability or knowledge of the individuals responding to those items. The difficulty of items is judged by the number of individuals who get them right, and the knowledge of individuals is judged by how many items they get right. Good test items discriminate among the ability levels of the test takers. For example, an item that only a few test takers answer correctly should be answered correctly mostly by individuals who scored well on the other items. Bad test items would be those that all test takers got right—or wrong—because these allow no distinctions in knowledge levels among the test takers. Another example of a bad item would be one that only high-ability test takers answered incorrectly.

The basic data used in Rasch modeling is the *ratio of the *probability of answering questions correctly to the probability of answering questions incorrectly. Since the *log of this ratio is used, the Rasch model is a kind of *logit model.

R

Rating Scale A *measurement technique for quantifying evaluations exemplified by the familiar "on a scale of 1 to 10 I'd give it a . . ." Because the person doing the rating is the measurement *instrument, and because rating scales usually involve assigning numbers to non-numerical phenomena, the process is often thought of as "subjective." But it can be quite *objective. Olympic judging is probably the best known example of how a high level of consensus can be reached about the number to be assigned to something non-numerical, such as a double back flip.

Ratio A combination of two numbers that shows their relative size. The relation between the numbers is expressed as a fraction, as a decimal, or simply by separating them with a colon. The ratio of one number to the other is that number divided by the other. The ratio of 12 to 6 (12:6) means 12/6, or 2:1. The statement "the ratio of X to Y" means "X divided by Y." Compare *proportion. See *odds ratio.

For example, say a birdwatcher counted the visits to his backyard feeder by different types of birds; if one afternoon there were 50 visits by nuthatches and 20 by chickadees, the ratio of nuthatches to chickadees could be expressed: 50/20, or 50:20, or 5:2, or 2.5:1.

Rationalism A philosophical doctrine which maintains that reasoning is the main way we get knowledge. This view is often contrasted with *empiricism, *positivism, and *realism, definition (b). The assumption of rationalism lives on in much economic theory and is at the heart of *game theory. *Postmodernism tends to reject both positivism and rationalism. See *empiricism, *idealism.

Rational Number An integer (a whole number), or a fraction composed of integers. Rational numbers can be expressed as *ratios; they are "ratio-nal." Compare *irrational number.

For example, 8,576, and −14 are rational numbers as are 4/9 and 13/27. But 85.76, 4.9, and 1.3/2.7 are not.

Ratio Scale (or Level of Measurement) A measurement or scale in which any two adjoining values are the same distance apart and in which there is a true zero point. The scale gets its name from the fact that one can make ratio statements about variables measured on a ratio scale. See *interval scale, *level of measurement.

For example, height measured in inches is measured on a ratio scale. This means that the size of the difference between being 60 and 61 inches tall is the same as between being 66 and 67 inches tall. And, because there is a true zero point, 70 inches is twice as tall as 35 inches (ratio of 2 to 1). The same kind of ratio statements cannot be made, for example, about measures on an *ordinal scale. The person who is second

R

tallest in a group is probably not twice as tall is the person who is fourth tallest.

Raw (Score or Data or Numbers)   Scores, data, or numbers that are in their original state ("raw") and have not been statistically manipulated. Note: The opposite of raw is not, in most cases, "cooked," since cooked implies dishonest manipulation of data. Usually, the expression "raw data" means data that need further work in order to be useful. Compare *derived statistics.

For example, if you got 23 out of 25 answers correct on a quiz, 23 would be your raw score. Other ways of reporting that score, such as 92% or 3rd in the class, would have to be calculated, and thus would not be "raw."

RCT   *Randomized clinical trial or *randomized control trial.

Reactive Measure   Any measurement or observation technique that can influence the subjects being measured or observed, as when surveying respondents about an issue influences their opinions about it. It is difficult to avoid such "reactivity." Reactivity is minimal when observing subjects in a *natural setting and in such a way that they are unaware that they are being observed. However, it is seldom possible to do such observing without becoming part of the setting and thus influencing the subjects indirectly. See *artifact, *Heisenberg principle.

Realism   (a) A philosophical doctrine holding that abstract *concepts really exist and are not, as *nominalism would have it, just names. Compare *holism, *methodological individualism, *latent variable. (b) A philosophical doctrine holding that the external world really exists, quite apart from our conceptions of it, and that the external world can be directly known by sensory experience. Often contrasted with *idealism. Compare *empiricism. Note that definitions (a) and (b) can be close to contradictory.

Real Limits (of a Number)   The points falling between half a measurement unit below and half a unit above the number.

For example, say you are measuring people's height to the nearest inch. The real limits of the height of someone whose height you record as 70 inches are 69.5 and 70.5; someone whose height falls between 69.5 and 70.5 will be recorded as being 70 inches tall.

Recall Data   Evidence provided by individuals based on their memories of past events. Virtually all data reported by individuals is recall data, but the term is usually reserved for data based on recollections of the relatively distant past. See *event history analysis.

Reciprocal (of a Number)   The reciprocal of a number is that number divided into 1 or (what amounts to the same thing) raised to the power of -1. For example, the reciprocal of 3 is $1/3$ or 3^{-1}; the reciprocal of X is $1/X$ or X^{-1}.

Reciprocal Relation   Said of a situation in which *variables can mutually influence one another, that is, can be both *cause and *effect. Compare *recursive.

For example, say that in an effort to lose weight you've taken up running. Running and weight loss could be reciprocal: The more you run the more you lose; the more you lose the easier running is, so you run more, which causes you to lose more, and so on.

Reciprocal Transformation   A *transformation of a data set by taking the *reciprocal of the members of the set.

Recode   See *coding and *transformation.

Record   A row in a computer matrix or *spreadsheet grid. See *case and, for an example, *field. Cases are recorded in rows; variables are in columns or fields.

Rectangular Distribution   Said of a distribution of a variable when its values do not change; when plotted on a graph, such a distribution forms a rectangle.

Rectilinear   (a) In a straight line; more often called *linear. (b) At right angles or perpendicular. See *orthogonal.

Recursive Model   A causal *model in which all the causal influences are assumed to work in one direction only, that is, they are *asymmetric (and the *error or disturbance terms are not correlated across equations). By contrast, nonrecursive models allow two-way causes. Contrast *reciprocal relation. Note: Readers are often confused by the term, because they take "recursive" to refer to recurring or repeating and therefore to mean a reciprocal relation.

For example, if you were looking at the influence of age and sex on math achievement, your model would probably be recursive since, while age and sex might influence students' math achievement, doing well or poorly in math certainly could not change their age or sex. On the other hand, your model of the relationship between achievement and time spent studying might not be recursive. It could be *reciprocal (or nonrecursive): Studying could boost math achievement, and increased achievement could make it more likely that a student would enjoy studying math.

Figure R.1 Recursive Model

Reductionism (a) A research procedure or theory that argues that the way to understand something is to reduce it to its smaller, more basic, individual parts—as organisms might be best understood by studying the chemicals of which they are made. Another example would be the view that the behavior of groups can only be understood by studying the behaviors of the individuals who make up the groups. The term reductionism is often pejorative; people who like such procedures more often call them *analysis or *methodological individualism. Compare *holism, *functionalism. (b) Reductionism can also be used to mean limiting explanations to a particular class of variables. For example, economic reductionism would be explaining everything by economic influences.

Redundant Predictor A variable in a *regression analysis that provides little or no additional ability to predict the value of the *dependent or *outcome variable. It is redundant because it is highly correlated with one or more other variables. Compare *confound, *intercorrelation, *multicollinearity.

Re-express To change, or *transform, values measured on one scale to another scale, as one might re-express speed in miles per hour as speed in kilometers per hour.

Reflexivity (a) Researchers' critical self-awareness of their biases and how these can influence their observations. (b) Difficulties of interpretation arising from the fact that researchers are almost always part of the context they are studying and thus cannot study it without influencing it. See *reactive measure.

Refusal Rate In survey research, the number of respondents who are contacted but refuse to participate in the survey; this number is divided by the initial sample size to get the rate. The non-response rate includes members of the sample who did not participate for any reason as well as those who refused. Refusers are usually only a relatively small portion of the non-response rate. See *non-response bias.

Region of Rejection An area in the tail(s) of a *sampling distribution for a *test statistic. It is determined by the *critical value(s) the researcher chooses for rejecting the *null hypothesis. Rejection regions are values of z or t distributions, not *probability (p) values. See the following illustration.

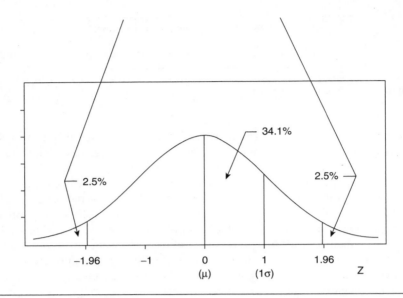

Figure R.2   Region of Rejection

Regressand   Another term for the *dependent, *outcome, or *response variable in a *regression analysis.

Regression   Any of several statistical techniques concerned with predicting or explaining the value of one or more *variables using information about the values of other variables. Regression is used to answer such questions as: "How much better can I predict the values of one variable, such as annual income (variously called the *dependent, *outcome, *criterion, *response, or Y variable), if I know the values of another variable, such as level of education (called the *independent, *predictor, *explanatory, or X variable)?"

The term "regression" originated in the work of the 19th-century researcher Francis Galton (1822–1911). In his studies of the heredity of characteristics such as height, he noted the phenomenon of *statistical regression, or regression toward the mean. Very tall people tend to have children somewhat shorter (closer to the mean) than themselves; and very short people tend to have children somewhat taller (closer to the mean) than they. Put more formally, extreme scores on a *predictor variable (parents' height) are likely to produce less extreme scores on a *criterion variable (children's height). Knowing how much regression toward the mean there is for scores on a particular pair of variables gives you a prediction. If there is very little regression, you can predict quite well. If there is a great deal

R

of regression, you can predict poorly if at all. The *r* used to symbolize *Pearson's correlation coefficient originally stood for "regression." Indeed, Pearson's *r* is the same thing as a standardized regression coefficient between two variables.

Regression Analysis (a) Methods of explaining or predicting the *variability of a *dependent variable using information about one or more *independent variables. Regression analysis attempts to answer the question: "What values in the dependent variable can we expect given certain values of the independent variable(s)?"

(b) Techniques for establishing a *regression equation. The equation indicates the nature and closeness of the relationship between two or more variables, specifically, the extent to which you can predict some by knowing others, that is, the extent to which some are associated with others. The equation is often represented by a *regression line, which is the straight line that comes closest to approximating a distribution of points in a *scatter diagram. See *regression plane.

Regression Artifact An artificial result due to *statistical regression or *regression toward the mean. Also called *regression effect.

Suppose a high school gave its students an English proficiency examination. The students with the very lowest scores were then assigned to a new tutorial program to help them prepare for the state-mandated graduation examination. If, after three months of tutoring, the students did better on a retake of the English proficiency examination, their improvement could be due to the effects of the program, but part of their improvement could also be due to regression artifacts. At least a few of the students probably got their original low scores through *random error (one wasn't feeling well the day of the test, another messed up the answer sheet, a third had her grade mistakenly recorded as a 68 when she earned an 86, and so on). Without a *control group, there would be no way to tell for sure how much of the improvement was due to regression artifacts and how much could be credited to the tutoring program.

Regression Coefficient A number indicating the values of a *dependent variable associated with the values of an *independent variable or variables. A regression coefficient is part of a *regression equation. A *standardized regression coefficient (one expressed in *z-scores) is symbolized by the Greek letter beta; an unstandardized regression coefficient is usually symbolized by a lowercase *b*.

For example, if we were studying the relation between education (independent variable) and annual income (dependent variable), and we found that for every year of education beyond the 10th grade the expected annual income went up by $1,200, the (unstandardized) regression coefficient would be $1,200.

Regression Constants In a *regression equation, the *slope(s) and the *intercept. In the typical formulation, $Y = a + bX$, a (the intercept) and b (the slope) are the constants; they are the same regardless of the values of the variables X and Y. "The constant," without specification, usually refers to the intercept.

Note that this usage, while common, is loose by the standards of mathematics, where a "constant" would have neither a variance nor a standard error, as do both the slope and the intercept.

Regression Diagnostics Various statistical techniques used to estimate the *reliability of *regression equations, often with emphasis on assessing the impact of *outliers. A particular focus is violations of key regression *assumptions, that is, the assumptions of *linearity, *equality of variances, and *normally distributed *errors.

Regression Discontinuity Design (RDD) A research design that uses a break in a regression line to indicate the presence of a treatment effect. It is used particularly when subjects are placed in control and treatment groups by a cutoff score (such as score on a diagnostic test) rather than by random assignment. It has been most widely used in the evaluation of educational programs.

Regression Effect The tendency in a pre-/posttest design for the posttest scores to regress toward the mean. See *statistical regression. Also called *regression artifact; see that entry for an example.

Regression Estimate The estimated value of the *dependent (or *outcome) variable in a regression equation after the values of the *independent (or *predictor) variables have been taken into account.

Regression Equation An algebraic equation expressing the relationship between two (or more) variables. Also called "prediction equation." Usually written $Y' = a + bX + e$. Y is the *dependent variable; X is the *independent variable; b is the *slope or *regression coefficient; a is the *intercept; and e is the *error term.

For example, in a study of the relationship between income and life expectancy we might find that: (1) people with no income have a life expectancy of 60 years; (2) each additional $10,000 in income, up to $100,000, adds two years to the average life expectancy so that people with incomes of $100,000 or more have a life expectancy of 80 years. This would yield the following regression equation: Life Expectancy = 60 years + 2 times the number of $10,000 units of income. $Y' = 60 + 2X$, where Y' equals predicted life expectancy, and X equals number of $10,000 units. The intercept is 60. (In this highly simplified example, there is no error term.)

Regression Line   A graphic representation of a *regression equation. It is the line drawn through the pattern of points on a *scatter diagram that best summarizes the relationship between the *dependent and *independent variables. It is most often computed by using the *ordinary least squares (OLS) criterion. When the regression line slopes down (from left to right) this indicates a *negative or inverse relationship; when it slopes up (as in the following illustration), this indicates a *positive or direct relationship. In this sample, when height is 1 inch greater, weight is about 9.1 pounds greater, on average, which means that the regression coefficient equals 9.1.

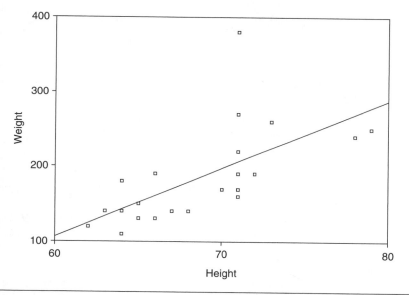

Figure R.3   Regression Line

Regression Model   Another term for *regression equation, that is, an equation postulating the relationship between a continuous *dependent variable, an *independent variable or variables, and an *error term. Regression modeling is a synonym for *regression analysis.

Regression Plane   A graphic representation of the relation of two *independent variables and one *dependent variable. When a regression has two independent variables, they can no longer be represented graphically by a line. They can, however, be represented by a two-dimensional plane in three-dimensional space, as in the following illustration.

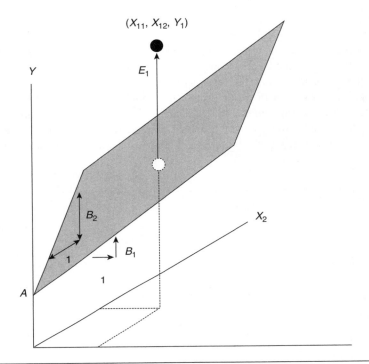

Figure R.4   Regression Plane

SOURCE: From John Fox, *Applied Regression Analysis, Linear Models, and Related Methods* (Thousand Oaks, CA: Sage, 1997), p. 98.

Regression SS   In *regression analysis, the sum of squares that is explained by the *regression equation. It is analogous to *between-groups sum of squares in *analysis of variance. Abbreviated: $SS_{regression}$. See *residual SS.

Regression Toward the Mean   A kind of *bias due to the fact that measures of a *dependent variable are never wholly reliable, which results in later measures of a variable tending to be closer to the mean than earlier measures. This tendency is especially strong for the most extreme values of the variable. Another term for *statistical regression. See *regression artifact for an example.

Regression Weight   Another term for *beta weight or *regression coefficient.

Regress On   In *regression analysis, to explain or predict by. To regress Y on X is to explain or predict Y by X. One regresses the *dependent variable on the *independent variable(s).

For example, the phrase "we will regress college grade point average (GPA) on SAT scores," means: We will try to explain (or predict) differences in students' GPAs by differences in their SATs.

Regressor Another term for *independent variable or *predictor variable. See *regress on.

Reification In *factor analysis and *principal components analysis, labeling the clusters of correlated variables discovered by the analysis (*latent variables) and treating them as though they were something concrete. As with reification more broadly, there is danger of overinterpretation when treating abstract theoretical entities as though they were substantive. Compare *realism.

Reinforcer Effect The tendency of one variable to increase a causal influence between or among other variables. The opposite is a *suppressor effect. Compare *intervening variable and *interaction effect.

In the following graphic, cognitive sophistication reinforces the causal link between education and tolerance.

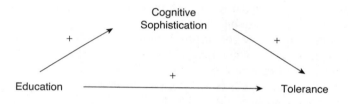

Figure R.5 Reinforcer Effect

Rejection Error Another term for *alpha error or *Type I error.

Relationship (or Relation) A connection between two or more *variables, usually assessed by a measure of *association, such as *Pearson's r. The connection can indicate a causal link or a mere statistical association.

Relative Frequency Another term for *proportion, that is, a number calculated by dividing the number of cases with a certain characteristic by the total number of cases. Relative frequency is often used as an estimate of *probability.

For example, if it rained on average 36.5 days per year, the relative frequency would be .1 (36.5/365). To get the percentage of days it rained, you would multiply the relative frequency by 100 (.1 × 100 = 10%).

Relative Frequency Distribution The proportion of the total number of cases observed at each score or value. For examples, see *frequency distribution, *probability distribution.

Relative Risk Ratio　See *risk ratio.

Relative Standing, Measure of　A statistic that describes the comparative position (or rank) of one case or subject to the others in a group. *Percentiles and *z-scores are two widely used examples.

Relative Variation, Coefficient of (CRV)　A measure of *dispersion used to compare across samples with different *means and *units of measurement. It is computed by dividing the *standard deviation of a *distribution by its *mean. Compare *z-score.

Relativism　A philosophical doctrine which maintains that there are no universal truths nor any objective knowledge. Knowledge claims are held to be relative to something about the persons making them, such as their culture or their vested interests. *Postmodernists are often accused of relativism; some of them deny it. See *cultural relativism.

Reliability　Freedom from measurement (or *random) error. In practice, this boils down to the consistency or stability of a measure or test or observation from one use to the next. When repeated measurements of the same thing give highly similar results, the measurement instrument is said to be reliable. Compare *validity.

　　For example, if you got on your bathroom scale and it read 145 pounds, got off and on again, and it read 139, repeated the process, and it read 148, your scale would not be very reliable. If, however, in a series of weighings you got the same answer (say, 145), your scale would be reliable—even if it were not accurate (*valid) and you really weighed 120.

Reliability Coefficient　A statistic indicating the *reliability of a scale, test, or other measurement. Reliability coefficients range from 0, when the measure is completely unreliable (that is, when all observed variance is due to *random error), to 1.0, when it is perfectly reliable. Scales or tests with coefficients less than .70 are usually considered unreliable. Most reliability coefficients are *correlations between two administrations, versions, or halves of the same test. See *Cronbach's alpha.

Repeated-Measures ANOVA　Analysis of variance in which subjects are measured more than once in order to determine whether statistically significant change has occurred from the *pretest to the *posttest. See *repeated-measures design. This research design goes by several different names, including within-subjects ANOVA and treatments-by-subjects ANOVA. See *randomized-blocks ANOVA.

Repeated-Measures Design　A research design in which subjects are measured two or more times on the *dependent variable. Rather than using different subjects for each *level of treatment, the subjects are given more than one treatment and are measured after each. This means that each subject is its own *control. See *correlated groups design, *panel study.

R

Replacement   See *sampling with replacement.

Replication   Said of research that tries to reproduce the findings of other investigators so as to increase confidence in (or refute) those findings. Repeating studies with different subjects or in different settings is especially important for *experimental laboratory research, because it helps increase *external validity. Despite the undeniable importance of replication, it is not done as often as it might be, largely because it is not considered high-status work. On the other hand, almost all studies that build upon past research are replication studies to some extent. Thus, replication is more a matter of degree rather than of kind.

For example, if one researcher found that people working in groups exerted less effort than when they worked individually, other researchers would be more likely to believe this finding if it could be replicated, particularly if it could be replicated using different kinds of subjects: women rather than men, adults rather than children, and so on.

Representative   Said of a *sample that is similar to the *population from which it was drawn. When a sample is representative, it can be used to make *inferences about the population. The most effective way to obtain a representative sample is to use *random methods to draw it. See also *probability sample.

For example, say you want to survey a sample of students on a college campus in order to draw conclusions about student opinion. You would have more confidence that your inferences were true of the student body as a whole if the sample strongly resembles the population. Let's say you know that 45% of all the college's students are male and 22% seniors. You could then look at your sample to see if there is a close match between your sample statistics and these known population *parameters. If the match is close, you can more confidently come to conclusions about the student body—more confidently, at any rate, than if your sample were, say, 80% male and 11% seniors. In that case your sample would not be representative; it would be *biased because it would overrepresent males and underrepresent seniors.

Representativeness   The extent to which a study's results can be generalized to other situations or settings. Another term for *external validity. See *representative.

Reproducibility   The tendency or ability to obtain similar results when *replicating a research study. Compare *reliability.

Resampling   The process of repeatedly sampling from a sample in order to estimate what other samples from a population would look like were we able to take them. Rather than taking numerous samples from a population to determine the *standard error (something that is usually done only in theory), one draws one sample and repeatedly resamples from it. This group of (re)samplings is then used for *confidence intervals and other *inferential statistics. The two best known techniques that use resampling are the *jackknife and *bootstrapping.

Research   Systematic investigation of a topic aimed at uncovering new information, solving a problem, and/or interpreting relations among the topic's parts. Research is done in hundreds of ways, ranging from lawyers searching among old court cases for legal precedents to physicists smashing atoms to study subatomic particles. While some research designs are clearly more effective than others for addressing particular sorts of problems, it is usually naïve to believe that there is one best way to conduct any research investigation.

Research and Development (R & D)   *Basic and *applied research aimed at inventing or designing new products. Thought of mostly in regard to manufacturing, perhaps especially to weapons manufacturing, R & D is increasingly common in service industries, such as banking and education, as well.

Research Design   The science (and art) of planning procedures for conducting studies so as to get the most valid findings. Called "design" for short. When designing a research study, one draws up a set of instructions for gathering evidence and for interpreting it. Compare *protocol.
    *Experiments, *quasi-experiments, *double-blind procedures, *surveys, *focus group interviews, and *content analysis are examples of types of research design.

Research Ethics   The principles defining good and bad conduct that (should) govern the activities of researchers. See *debriefing, *dehoaxing, *Milgram experiments, *subjects' rights, *IRB, *Nuremburg Code, *Helsinki Declaration.

Research Hypothesis   Often another term for the *alternative hypothesis. It is the hypothesis that one hopes indirectly to substantiate by rejecting the *null hypothesis. The alternative hypothesis is the logical opposite of the null hypothesis; the research hypothesis is not necessarily the same, although it often is. Symbolized H_1 or H_a.

Research Question   The problem to be investigated in a study stated in the form of a question. It is crucial for focusing the investigation at all stages, from the gathering through the analysis of evidence. A research question is usually more *exploratory than a *research hypothesis or a *null hypothesis.
    For example, a research question might be: What is the relation between A and B? A parallel research hypothesis could be: Increases in A lower the incidence of B. The associated null hypothesis might be: There is no

difference in the *mean levels of B among subjects who have different levels of A.

Research Review   Another term for *literature review; often used in conjunction with a *meta-analysis.

Research Strategy   A general plan for conducting research. *Experimental, *longitudinal, and *mixed-method are examples of research strategies. Some distinguish a research strategy from a *research design, which in this usage would be a more specific and detailed plan (tactic) for conducting research.

Residual   The portion of the score on a *dependent variable not explained by *independent variables. The residual is the difference between the value observed and the value predicted by a *model such as a *regression equation or a *path model. Residuals are, in brief, errors in prediction. "Error" in this context usually means degree of "inaccuracy" rather than a mistake. It is sometimes assumed that unmeasured *residual variables could account for the unexplained parts of the dependent variable. See *error term, *deviation score. In ANOVA designs, residual means *error variance or *within-groups variance. See *residual SS.

Residualize   To *control for or partial out.

Residualized On   Said of the variable that was *partialed out in a *semipartial correlation. Compare *regress on.
   For example, the semipartial correlation symbolized by $r_{1(2.3)}$ means the correlation between variables 1 and 2 after 3 was partialed out of 2—or after 2 has been residualized on 3.

Residual SS   In *regression analysis, the *sum of squares not explained by the *regression equation. Analogous to *within-groups SS in *analysis of variance. Also called "error SS."

Residual Term   Another expression for *error term, that is, the difference between an observed score or value and the score or value predicted by a model.

Residual Variable   In a *path analysis, an unmeasured variable; it may be assumed to cause the part of the *variance in the *dependent variable that is not explained by the path model.

Resistant Statistics   Measures less likely to be influenced by a few unusual, extreme values (*outliers) in a *distribution. Compare the *median (a resistant statistic) and the mean in the following two distributions. Increasing the last number from 7 (the value in Distribution X) to 77 (in Distribution Y) does not affect the median at all; but it more than triples the mean.

Table R.1   Resistant Statistics

	Median	Mean
X: 1, 2, 3, 4, 5, 6, 7	4	4
Y: 1, 2, 3, 4, 5, 6, 77	4	14

Respondent   A person who answers questions in a survey or an interview or otherwise responds to a researcher's inquiry.

Response Bias (or Effects)   Any biased outcomes due to interaction between interviewers and *respondents. For example, the ethnicity or sex of the interviewer might influence the way respondents answer questions.

Response Rate   The percentage or proportion of members of a sample who respond to a questionnaire. Low response rates are one of the more frequent sources of *bias in social science research.

Response Set (or Style)   (a) A tendency of subjects to give the same type of answer to all questions rather than answering questions based solely on their content. People who tend to say yes regardless of the question (yea-sayers) are a common example. (b) A tendency for observers to be influenced by an attitude toward the thing being observed. A prejudiced observer might, for example, rank some individuals lower than is merited by their actual behavior.

Response Variable   Another term for *dependent or *outcome variable; used most commonly in the context of *regression analysis.

Retrospective Study   Research that uses information from the past to draw conclusions. It is most widely used when the outcomes studied are long-term, which makes *longitudinal or *prospective studies impractical. *Case-control studies are a common example. For instance, if we were interested in the effects of early schooling on college graduation, we could conduct a prospective study by following students with different early school experiences to see whether and when they graduated from college years later. It would be easier and faster to conduct a retrospective study of graduates and nongraduates and compare their early schooling. See *recall data. Much research is in fact built on retrospective information, but the label is used mostly by those researchers whose data come from the distant rather than the more recent past.

Rho [P, ρ]   (a) A symbol for the *correlation coefficient for *population parameters; it corresponds to *Pearson's r, which is used for *sample statistics. (b) Abbreviation for *Spearman's (rank-ordered) correlation coefficient.

R

Ridge Regression   An alternative to *OLS methods of estimating regression coefficients. It is intended to reduce the problems in regression analysis associated with *multicollinearity.

Rise   The change in vertical distance between two points on a *regression line; it is used to calculate the *slope. See *run and, for a graphic, see *slope.

Risk   (a) The *odds or *probability of an unfavorable outcome, such as the value of your stock portfolio declining. Studies of risk usually involve statistical projections or modeling based on historical data. *Forecasts of the future are always uncertain, which is why acting on such predictions is risky. Calculations of risk are usually not simple probability estimates; they deal with probability and magnitude, such as your likelihood of losing money and, at each level of probability, how much you are likely to lose. Risk differs from pure uncertainty, where the probabilities of outcomes are completely unknown. See *hazard, *extrapolation. (b) The significance or *alpha level set by a researcher. This level is the researcher's statement of the acceptable level of risk of *Type I error.

Risk Ratio   A ratio of two probabilities of an unfavorable outcome that is used with *categorical variables. The risk ratio is very common in medical research. It is used for similar problems as the *odds ratio, which is more common in social science research. Although calculated on the same data and convertible into one another, the two ratios are not identical and are sometimes confused. The differences are most easily described using an example.

Say we took two random samples of 100 students each at a particular university to measure the graduation rates of males and females. The hypothetical samples are described in Table R.2. The male odds equal .67, which we get by dividing Yes by No, that is, 40/60 = .67. The female odds = 75/25 = 3.0. The odds ratio is: 3.0/.67 = 4.5, which means that the odds of females graduating are 4.5 times greater than the odds for males.

Table R.2   Risk Ratio: Graduation Rates for Two Samples of Students

	Yes/Graduate	No/Not Graduate
Males	40	60
Females	75	25

By contrast, risk ratios are based on probabilities, not odds. And they focus on the unfavorable outcome—in this case, not graduating. Be careful to remember that, although they are built on the same information and can be easily converted into one another, odds and odds ratios are *not* the same as probabilities and probability ratios. In our example, the *probability* of a

R

female not graduating is .25, that is, 25/100. The probability of a male not graduating is .60. The relative risk ratio (risk of not graduating) is: .60/.25 = 2.4. This means that the risk for males of not graduating is 2.4 times as great as the risk for females. The odds ratio, on the other hand, is 4.5; the odds of a male failing to graduate are 4.5 times as high. Which is the better way to report the relationship? The answer is found more in taste and tradition than in statistics. Odds ratios are used in *logistic regression and are more common in the social sciences. Probabilities and risk ratios are used in *probit regression and are more common in medical research. In both cases, the ratios range from zero to infinity, with 1.0 indicating equal risk or odds.

Robust   Said of a statistic that remains useful even when one (or more) of its *assumptions is violated—as long as the assumption is not violated too badly. Compare *resistant statistics.

    For example, the *F ratio (or ANOVA) is generally robust to violations of the assumption that *treatment groups have equal *variances, especially when the sample sizes are equal.

Robust Regression   Techniques used in *regression analysis to reduce the impact of *outliers on the estimates of regression coefficients. The usual method of calculating regression coefficients (*ordinary least squares or OLS) is, like all measures based on *means, susceptible to distortion in the presence of a few extreme values.

Role-Playing   In psychological investigations, researchers may assign roles for the research subjects to play, such as angry employer or defensive employee. Role-playing has many uses in therapy; in research it functions much the same way that computer simulations do. It enables the researcher to examine interactions in controlled and convenient contexts. *Game theory experiments often require the subjects to play roles. See *prisoners' dilemma.

Root   A number that results in a given number when it is raised to a given *power. See *radical.

    For example, 4 is the third root of 64, which means that $4^3 = 64$; also $\sqrt[3]{64} = 4$.

Root Mean Square (Error)   Another term for *standard deviation, so called because the standard deviation is calculated by taking the square root of the mean of the squared errors.

Rosenthal Effect   The influence of experimenters' expectations on outcomes of the experiment. The *Pygmalion effect is the best known, but probably not the most important, example. See *self-fulfilling prophecy, *demand characteristics.

Rotated Factor   See *factor rotation.

Rounding   Expressing numbers in shorter, more convenient units, that is, fewer decimal points than used when calculating the numbers.

For example, 13.834, 58.771, 61.213, and 97.098 could be rounded to 13.8, 58.8, 61.2, and 97.1.

Rounding Error   An error made by *rounding numbers before performing operations (adding, subtracting, etc.) on them.

The following table shows the number of bowls of soup served in a restaurant, and the percentage of each kind. "Percent (A)" shows the percentage of each flavor rounded to the nearest whole percent; but when totaled, this column adds up to 101%, a rounding error. To avoid that error, the restaurant owner would have to round to the nearest thousandth of a percent, as is done in "Percent (B)"—obviously, a silly level of precision in this case.

Table R.3   Rounding Error: Bowls of Soup Served, November 16–23

Kind of Soup	Number	Percent (A)	Percent (B)
Bean	46	14	13.855
Vegetable	72	22	21.687
Chicken Noodle	99	30	29.819
Tomato	115	35	34.639
Total	332	101	100.000

Row Marginals   The *frequency distribution of a variable shown across the rows of a *cross tabulation. See *marginal frequency distribution.

***r*-squared**   Another way of expressing "r^2."

R-Squared   Another way of expressing "R^2."

RSS   Regression *sum of squares, that is, the *explained variance. See *sum of squared errors (SSE).

***r*-to-Z Transformation**   A way of transforming *Pearson's *r* correlation coefficient that enables the researcher to compute a *confidence interval and *confidence limits for it. The transformation is done using natural logs. By using an *r*-to-Z transformation, the researcher can determine the *statistical significance of a Pearson's *r*. Note that the "z" is not a *z-score or standard score. Also called *Fisher's Z.

Rule   In statistics and mathematics, as in other areas of conduct, a rule is a statement that tells us what to do.

For example, the *formula A = B × C tells us that to find A, multiply B times C. A rule is often a formula stated in words rather than in symbols.

Run   (a) Repeated values of a variable; rolling dice and getting three 7s in a row would be a run. The *runs test is a *nonparametric test of the statistical significance of runs. (b) The change in horizontal distance between two points on a regression line. To get the *slope, divide the run into the *rise. See *slope for a graphic illustration.

Running Average   See *moving average.

Running Median   A way of *smoothing a line composed of *medians. The techniques are analogous to those used for *moving averages; and the purpose is the same, to make a trend clearer by reducing the visibility of fluctuations.

Runs Test   Another term for the *Wald-Wolfowitz test.

Ryan's Method   One of several ways of adjusting *significance levels to compensate for testing *multiple comparisons. See *Scheffé test, *Tukey's HSD, *Dunn test.

S (a) Abbreviation for *subject. Usually Ss, for subjects. (b) Lower- or uppercase italicized *s*, *standard deviation of a sample.

S^2 Lower- or uppercase italicized S^2, *variance of a sample.

Sample A group of *subjects or *cases selected from a larger group in the hope that studying this smaller group (the sample) will reveal important things about the larger group (the *population).

Sample Distribution A term sometimes used to refer to a *distribution of scores of a *variable in a *sample; an ordered array of the measurements or scores of a *sample of subjects. Not to be confused with a *sampl*ing* distribution.

  For example, if the Surgeon General were to conduct a study of the birth weights of children, a sample of all the live births in a given time period could be taken; the weights could be recorded and arranged in order, probably from heaviest to lightest, to facilitate study. This ordered arrangement would be the sample distribution and could be used, among other things, to calculate the relative frequencies of different weights.

Sample Point In *set theory, any member of the *sample space. Also called *elementary event.

Sample Size The number of subjects or cases selected for inclusion in a sample. Compare *sampling fraction.

Sample Size Formulas Several formulas exist for determining an appropriate sample size for a study. Many of these require information that one will not have until after the sample has been identified and studied. However, reviews of the literature can often enable researchers to make good guesses about that information. The general rule for sample size is the bigger the better, because the larger the sample the smaller the *standard errors. However, in a search for practical limits to sample size, researchers have used several formulas that help them identify samples that will be adequately large for their purposes.

For example, one suggestion for *multiple regression analysis is that this technique requires a minimum of 50 cases plus 8 cases for each independent variable. Other formulas are commonly used to calculate how big a sample needs to be to achieve a certain level of statistical *power.

Sample Space   A list of all the possible *samples of a given size that can be drawn from a particular *population. In *probability theory, a group of *data points (*elementary events) including all possible outcomes of an *experiment. See *underlying distribution.

For example, when talking about the probability of drawing particular cards from a deck, the deck of cards is the sample space. It contains all possible outcomes of any drawing.

Sample Survey   A survey in which a sample, not the whole *population, is studied.

Sample Weights   See *weighted sample.

Sampling   Selecting elements (subjects or other *units of analysis) from a *population in such a way that they are representative of the population. This is done to increase the likelihood of being able to generalize accurately about the population. Sampling is often a more accurate and efficient way to learn about a large population than a *census of the whole population. See *attrition.

Sampling Distribution (of a Statistic)   A *theoretical* *frequency distribution of the scores for or values of a *statistic, such as a *mean. Any statistic that can be computed for a sample has a sampling distribution. A sampling distribution is the distribution of statistics that *would be* produced in repeated *random sampling (with *replacement) from the same population. It is composed of all possible values of a statistic and their probabilities of occurring for a sample of a particular size.

A sampling distribution is constructed by assuming that an infinite number of samples of a given size have been drawn from a particular population and that their distributions have been recorded. Then the statistic, such as the mean, is computed for the scores of each of these hypothetical samples; *then* this infinite number of statistics is arranged in a distribution to arrive at the sampling distribution. The sampling distribution is compared with the actual *sample statistic to determine if that statistic is or is not likely to be the size it is due to chance.

It is hard to overestimate the importance of sampling distributions of statistics. The entire process of *inferential statistics (by which we move

S

from known information about samples to inferences about *populations) depends on sampling distributions. We use sampling distributions to calculate the probability that sample statistics could have occurred by chance, and thus to decide whether something that is true of a sample statistic is also likely to be true of a population parameter.

Sampling Error   The inaccuracies in inferences about a *population that come about because researchers have taken a *sample rather than studied the entire *population. Sampling error is an estimate of how a sample statistic is expected to differ from a population parameter in a *random sample drawn from the population. Also called *sampling variability. See *random error.

For example, to find out how many students at a particular college cheated on their schoolwork in the past year, you survey 200 (a sample) of them. Suppose you find that 28% said they had cheated. If your sampling procedure were a good one, 28% would probably be close to what you would have obtained had you surveyed all the students at the college. But your figure is likely to be off, say, as much as 2% one way or the other; that would mean that the true figure is somewhere between 26% and 30%. The sampling error is that figure plus-or-minus 2%. Compare *confidence interval.

Note that there are many other ways, besides sampling error, that your figure of 28% could be wrong. For instance, your questions could have been poorly worded or some students could have lied about whether or not they cheated.

Sampling Fraction   The size of a *sample as a percentage of the *population from which it was drawn; the *ratio of sample size to population size.

For example, a sample of 1,000 of the residents drawn from a town with a total population of 50,000 would yield a sampling fraction of 2% (1,000/50,000 = .02 = 2%). A sample of the same size in a city of a million would result in a sampling fraction of one tenth of one percent (1,000/1,000,000 = .001 = .1%). Sample size is usually more important than sampling fraction; our sample of 1,000 can be just as reliable for making inferences about the city of 1 million as about the town of 50,000.

Sampling Frame   A list or other record of the *population from which the *sampling units are drawn. It is an *operational definition of the population.

For example, suppose you wanted to select a sample of all the students at a university. If you picked every 25th name listed in the student telephone directory, the directory would be your sampling frame.

Note that a sampling frame might not include all members of the *population of interest (in fact, it almost never does). In the present example, some students do not have telephones, and others enrolled after the directory was published; they are part of the population, but they are not in the

S

sampling frame. Other students have graduated since the directory appeared; they are in the sampling frame, but are not part of the population.

Sampling Units Items from a *population selected for inclusion in a *sample. Compare *unit of analysis.

For example, if the population were all North American cities with more than 5,000 residents, and the sample were 200 such cities, they would be the sampling units.

Sampling Variability (or Variation) (a) Differences in a *statistic when it is computed on two or more *samples drawn from the same *population. (b) Differences between a statistic computed for a particular sample and that statistic computed for the population. See *sampling error.

Sampling With (or Without) Replacement When drawing a sample from a *population, one can replace or not replace subjects into the *sampling space after each draw. Many statistical procedures assume sampling with replacement, and this is the preferred method when possible. However, for practical reasons, most samples actually used in the social and behavioral sciences are drawn without replacement. When the population or sample space is very large, this violation of statistical assumptions is seldom serious.

For example, if you drew cards (sampled) one at a time from a deck and recorded the suit of each, you could replace the card and reshuffle after each draw or not do so. The procedure matters. If you got a diamond on the first draw, replaced it, and reshuffled, your chances of getting a diamond (with replacement) would be the same on the second draw as they were on the first (13 out of 52, or 25%). On the other hand, without replacement after the first draw there would be one less card in the deck (the diamond). That would reduce your chances of drawing a diamond on the second draw (to 12 out of 51, or 23.5%).

When sampling with replacement, each draw of the cards (event) is *independent of every other draw. When sampling without replacement, each draw is *conditional on the previous draws.

SARIMA Seasonally adjusted *ARIMA.

SAS Statistical Analysis System. A widely used *statistical package for data analysis in the social and behavioral sciences. Compare *SPSS, *BMDP.

Saturated Model A model that takes into account all possible effects, including interaction effects, of the variables in a study. This is done by having as many variables as there are values to estimate. For example, in a *factor analysis, a saturated model would calculate a number of factors equal to the number of variables. One can judge the value of a simpler (unsaturated) model, one with fewer factors, by comparing it to the saturated model.

S

Scalar (a) Like or pertaining to a *scale. Said of questionnaire items that can be arranged in a definite logical order. See *Guttman scale. (b) In *matrix algebra, a scalar is an ordinary number, not a matrix or a *vector. For example, when each component or element of a matrix is multiplied by a single number, this is called scalar multiplication.

Scale (a) A set of numbers or other symbols used to designate characteristics of a variable used in *measurement. For example, the numbers on a thermometer and the words "low, medium, and high" on an air conditioner are scales. See *level of measurement. (b) A group of related measures of a *variable. The items in a scale are arranged in some order of intensity or importance. A scale differs from an *index in that the items in an index need not be in a particular order, and each item usually has the same weight or importance. For examples see the *Bogardus, *Guttman, *Likert, and *Thurstone scales. Many writers do not distinguish between a scale and an index; it is fairly common to use the terms interchangeably to refer to any composite measure or *summated scale. Other writers consider this to be sloppy usage.

Scale Attenuation Effect See *floor and *ceiling effects.

Scale of Measurement Another term for *level of measurement.

Scale Up To take research findings and apply them more broadly. For example, findings from a learning laboratory might be used to guide statewide education reform, or discoveries in a chemistry lab could be scaled up to an industrial application.

Scaling (a) Another term for *measuring. See *nominal, *ordinal, *interval, and *ratio scales. (b) The process of creating a *scale by putting a group of related items in a logical sequence.

Scatter Diagram See *scatter plot.

Scattergram See *scatter plot.

Scatter Plot Also called "scatter diagram" and "scattergram." The pattern of points that results from plotting two *variables on a graph. Each point or dot represents one *subject or *unit of analysis and is formed by the intersection of the values of the two variables. See *data point.

The pattern of the points indicates the strength and direction of the *correlation between the two variables. The more the points tend to cluster around a straight line, the stronger the relation (the higher the correlation). If the line around which the points tend to cluster runs from lower left to upper right, the relation is *positive (or direct); if it runs from upper right to lower left, the relation is *negative (or inverse). If the dots are scattered randomly throughout the grid, there is no relationship between the two variables. For examples, see *correlation coefficient, *heteroscedasticity, *intercept, and *regression line.

Scedasticity　The degree to which the values of a *dependent variable are scattered or dispersed across the values of an *independent variable. The question analysis of scedasticity tries to answer is: As the values of the independent variable change, does the degree of *dispersion in the dependent variable's values remain uniform? See *homoscedasticity and *heteroscedasticity for a graphic illustration.

Scenario　An assumed or imagined sequence of events used to make decisions and contingency plans about future actions or events—often by contrasting two or more scenarios. Compare *decision table, *model.

Schedule　See *interview schedule.

Scheffé Test　A test of *statistical significance used for *post hoc multiple comparisons of means in an *ANOVA. Among its main features are that it is a conservative test (it tends to err on the side of underestimating significance) and that it deals well with unequal *cell sizes. The Scheffé test involves calculating and using a new, more demanding *critical value for F. See *omnibus test.

Schema　(plural: schemata) Generally, a diagram, plan, or framework. In cognitive psychology, philosophy of science, and related fields, a system for codifying concepts, experience, and data that organizes the way we perceive, learn, and remember. Compare *paradigm, *model.

Science　Most generally, the discovery, creation, accumulation, and refinement of knowledge. There are many controversies about what science is and about which disciplines, specialties within disciplines, and methods are "truly" scientific. See *epistemology, *paradigm.

Scientific Hypothesis　Another term for *alternative hypothesis. Also called *research hypothesis. Compare *null hypothesis, *research question.

Scientific Significance　Often contrasted with *statistical significance, scientific significance refers to the degree that a finding helps advance knowledge in a field. Scientific significance is decided by judgment, not statistical routines. See also *practical significance, *substantive significance, *clinical significance.

Scientism　A term used to belittle or dismiss theories and people who tend to treat science as a religion or an ideology.

　　For example, people who believe that science is the highest human value, that it can solve all real problems, and that it is superior to any other form of knowledge or belief are likely to be accused of scientism. Compare *positivism.

Scope Conditions   The specifications or limitations on the applicability or *validity of a *theory and its *propositions. Scope conditions indicate the circumstances (such as times, places, kinds of subjects) under which a theory's propositions would hold true.

Score   A value of a variable. Used both as "score" in the ordinary sense of the term, such as a test score, and more broadly to mean a measurement or value. See z-score.

Scree Plot   A plot of *eigenvalues in descending order. So called after its resemblance to a geologic scree, which is an accumulation of stones or other rubble at the base of a steep hill. Eigenvalues high up on the "hill" would be included as factors worthy of further analysis; those at the bottom would not. The division between the included and the excluded is the point where the eigenvalues level off.

The example shows the scree plot for the eigenvalues of 37 variables. Visually, the leveling-off point is at the fourth factor (actual eigenvalue = 1.8).

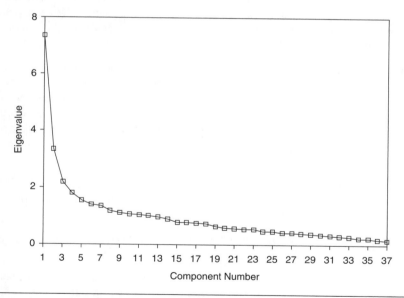

Figure S.1   Scree Plot

SD   (a) *Standard deviation (italicized). Also written *Sd* or *sd*. (b) *Semantic differential (not italicized).

SDT   *Signal detection theory.

SE *Standard error, often standard error of the mean.

Seasonally Adjusted Said of time-series data when regular, seasonal fluctuations are statistically removed. For example, if unemployment rates have regularly increased by 1% in the winter months, an increase in the winter months would have to exceed 1% before it would be interpreted as a true increase.

Secondary Analysis A type of research in which *data collected by others are reanalyzed. In some fields (sociology is one; psychology isn't), secondary analysis may be the most common form of research.

Examples of extensive and widely used data *archives available for secondary analysis include the U.S. Census *Public Use Microdata Samples (PUMS), the *National Election Survey (NES) of the Inter-University Consortium for Political and Social Research, and the *General Social Survey (GSS) of the National Opinion Research Center (NORC). Secondary analysis in qualitative research can be thought of as having a long tradition (e.g., historians using documents in archives) or as emerging quite recently (e.g., publicly available oral history transcripts).

Secondary Data Information collected by other researchers but available, sometimes in *databases, for use by others. Huge amounts of such data are available, for example, at the National Institutes of Health and the National Center for Education Statistics. See *secondary analysis for other examples.

Secondary Source A source that provides non-original ("secondhand") data or information. Compare *primary source, *original research.

For example, if you wanted to study the city council, you could interview its members and attend its meetings; or you could read newspaper articles by a reporter who had interviewed the members and attended their meetings. In the latter case you would be relying on a secondary source—you would not have obtained the information firsthand.

Second-Order Interaction An interaction among three *independent variables and a dependent variable. A first-order, or simple, interaction is between two independent variables.

Second-Order Factor Analysis Studying *correlations among factors so as to find factors that lie behind factors. See *factor analysis.

Second-Order Partial Said of a *correlation or *regression coefficient when two variables are controlled. See *higher order partials.

Secular Trend A long-term trend or one of indefinite length, usually as opposed to a short-term trend or fluctuation. See *moving average, *time-series data.

Segmented Bar Chart A bar chart in which each of the bars is made up of two or more parts ("segments"). This allows you to use a bar chart to compare both wholes and their parts. Also called "stacked" bar chart.

The following examples show the number of deaths per 100,000 attributable to heart disease for selected years for the whole U.S. population. The segmented bars and the clustered bars show the separate rates for men and women. While one or the other approach to graphing could be better for different purposes, the choice is mostly a matter of taste.

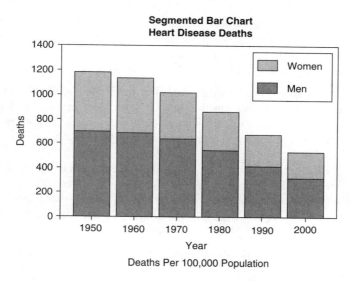

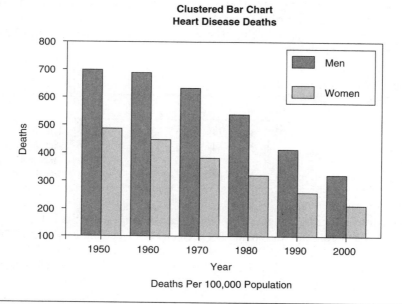

Figure S.2   Segmented Bar Chart (Compared to Clustered)

Selection Effect See *self-selection bias.

Selection Threat See *self-selection bias.

Self-Fulfilling Prophecy Something that happens because people expect it to happen. (The term was coined by Robert K. Merton.) Compare *Pygmalion effect.

For example, suppose account holders believed that Billy-Bob Savings and Loan was about to go bankrupt because depositors would soon be withdrawing all their money. If enough depositors predicted this, it could happen—even if B-B S&L were a financially sound institution.

Self-Selection Bias Also called "selection threat (to validity)" and "self-selection effect." A problem that may arise in the comparison of groups when the groups are formed by individuals who choose to join them and thus are not formed by a researcher assigning subjects to *control and *experimental groups.

For example, comparing the effects on academic achievement of attending two-year colleges versus four-year colleges would be difficult because (among other reasons) students who chose to attend two-year colleges might be different in important ways (e.g., goals, income, motivation, aptitude) from those who selected four-year colleges.

SEM (a) *Standard error of the mean. (b) *Structural equation modeling.

Semantic Differential Scale A question format in an interview or a survey in which *respondents are asked to locate their attitudes on a scale ranging between opposite positions on a particular issue (described by bipolar adjectives), as in the following. The subject circles a number from 1 to 7 for each scale.

Table S.1   Semantic Differential Scale

My Boss Is							
Fair	1	2	3	4	5	6	7	Unfair
Passive	1	2	3	4	5	6	7	Active
Lazy	1	2	3	4	5	6	7	Hardworking
Efficient	1	2	3	4	5	6	7	Inefficient

Semi-Interquartile Range The *interquartile range (IQR) divided by 2. In other terms, the difference between the 75th percentile and the 25th percentile divided by 2. Also called *quartile deviation.

For example, in the series 4 6 8 10, the range is $10 - 4 = 6$; the IQR is $8 - 6 = 2$; the semi IQR is $2/2 = 1$.

Semi-Logarithmic Ruling Said of graph paper using *logarithmic ruling on only one *axis (usually the vertical or y axis). See *logarithm. Such graph paper is often used for *time-series studies to compare trends for sets of data, especially when the sets of data are of different orders of magnitude, such as the populations of a city and of a nation.

Semiology The study of signs and their meanings. A branch of linguistics and linguistic philosophy that some theorists believe has great promise as a model for how to study several aspects of human thought and action.

Semipartial Correlation A correlation that partials out (*controls for) a variable, but only from one of the other variables being correlated. Also called "part correlation." See *partial correlation, which partials out a variable from all other variables. It is computed using *multiple regression analysis.

Sensitivity The ability of a diagnostic test to correctly identify the presence of a disease or condition. Sensitivity is the *conditional probability of the test giving a positive result if the subject does have the condition or disease. Originating in medical research, the term is now used more broadly. Compare *power of a test, *specificity.

Sequential Analysis (a) A kind of investigation in which researchers decide as they go along (through periodic analyses of the data gathered up to that point) how much more and what kinds of data should be gathered next. Compare *negative case analysis, *post hoc comparison. (b) Descriptions and analyses of the sequences in which behavior occurs; often conducted using data gathered from detailed observations of behavior in natural settings.

Sequential Sampling Observations made until enough data have been gathered to make a decision. This contrasts with the more typical procedure of deciding on the number of observations (or sample size) before beginning the study. Compare *Pascal distribution.

Serial Correlation Another term for *autocorrelation, that is, correlation between earlier and later items in a *time series.

Serial Dependency This occurs when behavior at later observations is dependent upon behavior at earlier observations. Compare *A-B-A-B design, *repeated measures design.

SES *Socioeconomic status.

Set A well-defined group of things. Events, objects, or numbers that are distinguishable from all other events, objects, or numbers on the basis of some specific characteristic or rule.

S

Examples of sets include: all even numbers (those that are exactly divisible by 2); all murders (intentional illegal killings of persons, as opposed to all other forms of death); all female physicians (contrasted with male physicians and/or females who are not physicians). Compare *sample space, *fuzzy set.

Set-Theoretic Model   An application of *set theory to the analysis of data.

Set Theory   A branch of logic and mathematics that deals with the characteristics of and relations among *sets.

Shapiro-Wilks Test   A statistical test of the *hypothesis that sample data have been drawn from a population with a *normal distribution. It can also be used to test for other population distributions. Usually abbreviated W. Compare *Lilliefors test and *normal probability plot.

Shotgun Approach   Said of research in which the investigator studies a large number of variables with no clear strategy or theoretical justification. The researcher points, so to speak, in a general direction and hopes to hit something. Compare *fishing expedition.

Shrinkage   The tendency for the strength of prediction in a *regression or *correlation study to decrease in subsequent studies with new data sets. The regression model derived from one set of data usually works less well with others. The degree of shrinkage is measured by change in R^2. Compare *regression to the mean.

Sigma [Σ, σ]   (a) The uppercase sigma usually means "sum of," and is thus an indication that the numbers following it are to be (or have been) added together. (b) Lowercase sigma is often used to symbolize the *standard deviation of a *population. See *standard score. (c) Lowercase sigma squared means population *variance.

Sigmoid Curve   A curve resembling the letter S.

Signal Detection Theory (SDT)   The theory that the detection of a signal, such as a sound, depends on sensory input and a decision process about whether and what one has heard. It is interesting for methodology because of the parallels between the decision process of the subject in SDT and the researcher in *hypothesis testing. When "signal" is interpreted broadly to include any outcome that can be present or not, such as a symptom of a disease, SDT can be used in many areas of research. SDT is not different in principle from *hypothesis testing, but most people find SDT easier to understand than the double and triple negatives used to describe hypothesis testing.

The following matrix shows the possible outcomes in a signal detection experiment. Compare this with the similar illustration of hypothesis testing.

Table S.2   Signal Detection Theory

		Subject's Decision About Signal	
		Yes	*No*
Actual	Yes	Hit	Miss
Presence	No	False Alarm	Correct Rejection
of Signal			

Significance   The degree to which a research finding is meaningful or important. Without qualification, the term usually means *statistical significance, but lack of specificity leads to confusion (or allows obfuscation). See *practical significance, *substantive significance, *significance testing.

Significance Level   The *probability of making a *Type I error, that is, the probability at which it is decided that the *null hypothesis will be rejected. The lower the probability, the greater the *statistical significance. Also called *alpha level. See *p value.

Significance Testing   Using statistical tests (such as the *chi-squared test, *t-test, * z-test, or *F test) to determine how likely it is that observed characteristics of a *sample have occurred by chance (or *sampling error). If the observed characteristics in the samples are unlikely to have been due to chance, the characteristics are deemed statistically significant and can be used to make inferences about *populations. Compare *hypothesis testing. See *sampling distribution.

Sign Test   The simplest and oldest of all *nonparametric statistical tests. So called because the statistic is computed from data in the form of plus and minus (+ and −) signs.

For example, in a two-group experiment, the researcher could simply assign a + to each case where the score in Group A was higher than that in Group B and a − when the reverse was true. The *null hypothesis for a sign test is that there is no difference in the number of pluses and minuses.

Simple Classical Probability   Probability statements about a *sample space whose *elementary events are all equally likely to occur. See simple *random sample.

For example, the sample space of an evenly balanced coin contains the two elementary events "heads" and "tails." Making statements about the likelihood of getting a particular ratio of heads to tails in a given number of tosses would be an exercise in simple classical probability.

Simple Correlation   A *correlation between two variables only or a correlation that describes a *linear relationship. Used in contrast with a *multiple

S

correlation or with a *nonlinear relation—both of which are more complex than a simple correlation.

Simple Random Sampling   See *random sampling.

Simple Regression   A form of *regression analysis in which the values of a *dependent variable are attributed to (are a function of) a single *independent variable. Usually contrasted with *multiple regression analysis, in which two or more independent variables are used to explain one dependent variable.

Simpson's Paradox   A surprising or contradictory result that occurs when a relation between two variables is reversed when taking a third variable into account—or when the direction of the values for two groups taken individually is reversed when the groups are merged. *Lord's paradox is a variant. See *ecological fallacy.

Table S.3 illustrates. It gives the scores of two classes of 10 students in two different years. Students are in two groups, A and B, one of which scores substantially higher than the other. In the first year, the five students in Group A score a mean of 80 and the five in Group B get a mean of 60. In the second year, both groups raise their averages somewhat, to 81 and 62. And the gap between the groups narrows slightly, but the overall average goes down. That's the paradox: Both groups do better, but the overall mean goes down. The reason is that in the second year there are two fewer students in the high-scoring group and two more in the low-scoring group.

Table S.3   Simpson's Paradox: Scores of Students

	Year 1		Year 2	
	Group A	Group B	Group A	Group B
	80	60	81	62
	78	65	79	67
	82	55	83	57
	75	58		64
	85	62		60
				63
				61
Group Means	80	60	81	62
Annual Means		70		67.7

Simulation   See *computer simulation, *Monte Carlo methods, *role-playing.

Simultaneous Equations   A problem in *regression analysis that occurs when a variable can be both a cause and an effect. Alternatives to *least squares methods must be used when the direction of causation is unclear or when some variables are believed to be both the cause and effect of others. If we thought that X caused Y and that Y caused X, to understand the whole we would need to look at two equations simultaneously, one for how X caused Y and one for how Y caused X. If the regression's *predictor variables are *latent variables, then *structural equation modeling is used. In brief, when reality is more complicated than can be modeled with one equation, multiple equations considered simultaneously may have to be used.

For example, if the number of crimes in a community goes up, the number of police may tend to be increased. But, as the number of police goes up, the number of crimes may tend to go down. Or, to take a second example, a drop in the level of rainfall can lead to the spread of the desert, which can lead to further declines in rainfall. In both cases we would need to use simultaneous equations for a regression analysis.

Single-Blind   A study design in which the subject or patient is kept ignorant of the treatment, but the researcher is not. See *double-blind.

Single-Classification ANOVA   Another term for *one-way ANOVA. Also called "single-factor ANOVA."

Single-Factor ANOVA   Another term for *one-way ANOVA. Also called "single-classification ANOVA."

Six Sigma   A group of quality control measures that aims to insure that there are at least six *standard deviations (sigmas) between the mean and the nearest specification limit. This translates into a tolerance for error of less than 4 per million.

Skewed   Sometimes used casually to mean unusual, odd, or nonrepresentative. See *skewed distribution.

Skewed Distribution   A distribution of scores or measures that, when plotted on a graph, produce an asymmetrical curve. In a unimodal skewed *frequency distribution, the *mode, *mean, and *median are different. When the skewness of a group of values is zero, their distribution is symmetrical.

A positively (or upward or right) skewed distribution is one in which the infrequent scores are on the high or right side of the *x axis, such as the scores on a difficult test. A left (or downward or negatively skewed) distribution is one in which the rare values are on the low or left side of the x axis, such as the scores on an easy test. One way to sort out which is which is to remember that a skewer is a pointy thing; when the pointy end of the distribution is on the right, it is right skewed, and conversely for left skewed. Compare *normal distribution, *kurtosis.

S

(Left, Downward, or Negative Skew)

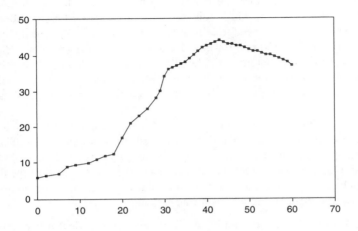

(Right, Upward, or Positive Skew)

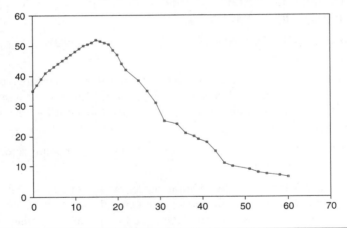

Figure S.3   Skewed Distribution

Skewness   The degree to which measures or scores are bunched on one side of a *central tendency and trail out (become pointy, like a skewer) on the other. The more skewness in a *distribution, the more *variability in the scores. See *skewed distribution for illustrations. Compare *kurtosis.

Computer programs often compute indexes of skewness. Positive values indicate a positive or right skew. Negative values indicate a negative or left skew. The value for a *normal distribution is zero.

Sleeper Effect   Originally used in mass communications theory to refer to a delayed reaction to propaganda or other message. More generally, any effect that becomes apparent only after the passage of time.

Slope   Generally, the rate at which a line or curve rises or falls when covering a given horizontal distance. The most common reference is to the steepness or angle of a *regression line, usually symbolized by the letter b in a *regression equation. The slope is calculated by taking any two points on the line and dividing the vertical distance (the "rise") between them by the horizontal distance (the "run") between them.

In a *positive or *direct relationship, the line slopes upward from left to right (see the following illustration). In a *negative or *inverse relation, it goes down from left to right (called negative because the slope b is a negative number).

Positive and Negative Slopes (Run and Rise)

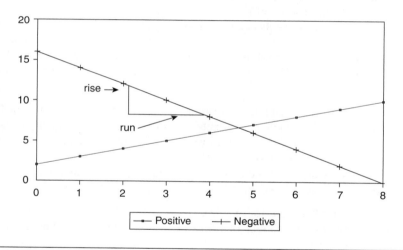

Figure S.4   Slope

Slope Analysis   Another term for *regression analysis, so called because it focuses on the angle or slope of a *regression line, the steepness of which indicates the strength of a relation.

Slutsky-Yule Effect   When *outliers occur in a *time series to which *moving average techniques have been applied, this can result in misleading fluctuations, especially when the outliers are random. Such changes in trend lines due to random outliers are called the Slutsky-Yule effect.

S

Smoothing Reducing irregularities (*fluctuations) in *time-series data, generally by using a *moving average. This is done to make long-term trends more apparent. See *ARIMA and, for an illustration, *moving average.

SMSA *Standard Metropolitan Statistical Area. Also called MSA for short.

Snowball Sampling A technique for finding research subjects. One subject gives the researcher the name of another subject, who in turn provides the name of a third, and so on. This is an especially useful technique when the researcher wants to contact people with unusual experiences or characteristics who are likely to know one another—members of a small religious group, for example. Also called *network sampling.

Social Constructionism See *constructionism.

Social Desirability Bias Bias in the results of interviews or surveys that comes from subjects trying to answer questions as "good" people "should" rather than in a way that reveals what they actually believe or feel. Because social desirability bias is very difficult to study, there is little reliable evidence about its extent. Some researchers believe it is a big problem, others do not.

For example, if you asked people, "Are you a racist?" most would probably say, "No." Hardly anyone thinks it is acceptable (socially desirable) to admit to being a racist, including people who are racists, at least by some definitions.

Social Indicators *Statistics describing *variables that reflect social conditions, that is, that "indicate" something about the nature and quality of life in a society, often as it changes over time. The term is used by the U.S. Census Bureau as well as other researchers.

Examples of social indicators include per capita income, average life expectancy, median years of education, and infant mortality rate.

Social Science Citation Index (SSCI) A multivolume source, compiled since 1966, of the citations of works in articles published in thousands of scholarly journals.

The SSCI has many uses. For example, if you found an excellent article published in 1990 on your topic, you could use the SSCI to see where that article had been cited in the years since it was published. This would be one way to learn what work has been done on your topic since 1990.

Social Sciences Any of several areas of study that focus on human interaction, institutions, and culture. Social sciences include sociology, economics, psychology, anthropology, political science, history, and some aspects of geography.

Socioeconomic Status (SES) Any of several composite measures of social rank usually including income, education, and occupational prestige.

Sociogram Also called "sociograph." A graphic representation of the relations among individuals in a group—constructed on the basis of some *sociometric measure. See *network analysis.

For example, the answers to the question, "Who would you most like to work with?" for six people (see the example at *sociometric matrix) could be graphed as follows:

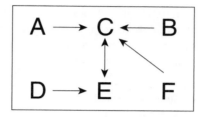

Figure S.5 Sociogram

Sociometric Matrix A rectangular arrangement of *sociometric measures. Compare *sociogram.

For example, suppose that a manager asked the six members of one of her production departments two questions: With which member of your group would you most like to work closely on a special project over the next year? Which one would you least like to work closely with? Answers could be useful in planning work assignments.

The results might look like those in the following matrix. 1 = "most like to work with" and −1 = "least like to work with." The matrix can be read vertically (by columns) or horizontally (by rows). Reading horizontally, we see, for example, that A would most like to work with C and least like to work with B. Reading vertically, it becomes clear that A's choices are fairly typical. Most people choose C for "most like to," and most choose B for "least like to." We can also see that no one has strong feelings one way or the other about A or F, and that opinions are evenly split about E.

Table S.4   Sociometric Matrix

	A	B	C	D	E	F
A	x	−1	1			
B		x	1		−1	
C			x	−1	1	
D		−1		x	1	
E		−1	1		x	
F			1		−1	x

Sociometric Measure   Any of several ways to find out about the existence and the strength of relationships among individuals in a group.

    For example, grade school students might be given a *questionnaire with items such as: "Who is your best friend in the class? Who do you play with at recess?" Alternatively, similar *data could be gathered more slowly, but perhaps yielding results of greater *validity, by observing with whom children play at recess.

Sociometry   Any of a number of methods of gathering data on the attractions, repulsions, interactions, communications, and choices of individuals in groups. Most often called *network analysis today. See *sociogram, *sociometric measure, *sociometric matrix.

Software   Instructions, or *programs, that tell a computer how to perform tasks. Software is often contrasted with "hardware," which refers to the physical computer itself. Word processing, spreadsheet, and graphics programs are examples of software.

Solipsism   The philosophical doctrine that denies there is any reality outside oneself; also, the belief that one can know nothing other than one's own reflections. *Relativism is often accused of leading to solipsism.

Solomon Four-Group Design   A type of *factorial design that helps researchers *control for or measure *pretest sensitization. Subjects are randomly assigned to one of four groups. These receive different combinations of the pretest, treatment, and posttest as illustrated in the following table.

Table S.5   Solomon Four-Group Design

	Pretest	Treatment	Posttest
Group 1	Yes	Yes	Yes
Group 2	Yes	No	Yes
Group 3	No	Yes	Yes
Group 4	No	No	Yes

Somers's d   An *asymmetric *measure of association for variables measured on an *ordinal scale. It is a *PRE measure.

Span   The difference between the lowest and highest values in a distribution of values. More often called *range.

Sparse Table   A table with many empty *cells or with many frequencies of zero.

Spearman-Brown Formula   Used to predict the approximate gain in the *reliability with which something could be measured by increasing the number of observations. Often used to adjust (upwards) a *split-half reliability estimate.

Spearman Correlation Coefficient (Rho)   A statistic that shows the degree of *monotonic relationship between two *variables that are arranged in rank order (measured on an *ordinal scale). Also called "rank-difference correlation." Abbreviated r_s for a sample. See *correlation coefficient, *Pearson correlation coefficient, *Kendall's tau.

For example, suppose you wanted to see if there was a relationship between knowledge of the political system and self-esteem among college students. You take a *sample of students and give them a test of political knowledge and a psychological assessment of their self-esteem. Then you rank each of the students on the two scales. Spearman's rho measures the association between the two sets of ranks. The *null hypothesis is that the two ranks are *independent.

Specification   Stating the propositions of a theory or model; saying (specifying) that a relation exists and indicating under what conditions it will be larger or smaller, such as describing how changes in the independent variable will result in changes in the dependent variable. In *path analysis, specification is drawing the path diagram. See *specification error.

Specification Error   A mistake made when deciding upon (specifying) the *causal model in a *regression analysis, or *path analysis, or *structural equation model. Common specification errors include leaving an important variable out of the causal model and including an irrelevant variable.

It can be very difficult to tell when such an error has been made, because much of the purpose of regression analysis is to decide which variables are important and which are irrelevant. In other words, you have to know which variables are important, and include them in the model, in order to determine whether they are important. See *paradox of inquiry.

Specification Problem   The problem of how to decide (how to specify) which variables to include and which to exclude in a *regression equation and, in a *path analysis, the direction of the causal arrows. See *specification error.

Specificity   The ability of a test to correctly judge that subjects do not have a disease or condition, in other words, to avoid *false negatives. Specificity is the *conditional probability of a test giving a negative result when patients or subjects do not have a disease. Originating in medical research, the term is used more broadly today. Compare *sensitivity, which is the ability of a test to avoid *false positives.

Sphericity Assumption   A statistical assumption important for *repeated-measures ANOVAs and *multiple regression. The assumption is that, at different levels of the *independent variables, the errors are uncorrelated and have equal variance. See *homoscedasticity. When it is violated, *F values will be positively biased. Researchers adjust for this bias by raising the *critical value of *F* needed to attain statistical significance. *Mauchley's test for sphericity is the most common way to see whether the assumption has been met.

Spike   In *time-series data, a sharp increase followed by a quick decrease in a variable. See illustration, in which the spike occurs at year 9.

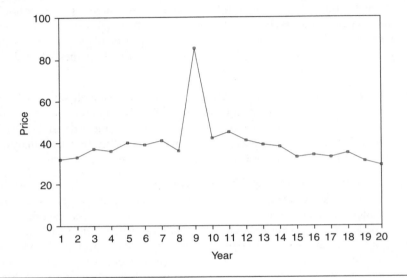

Figure S.6   Spike

Spline Regression   A technique used when the regression line, especially one representing change over time, contains bends or turns; in other words, when the regression line is not *linear. The points at which turns occur are called "spline knots." The main alternative, *polynomial regression, is less effective because it is more vulnerable to *collinearity problems.

Split-Half Reliability   A way to check the *internal consistency (or *reliability) of an *index by seeing how well the scores on one half of the items *correlate with those on the other half. Compare *Cronbach's alpha, *KR20.

For example, suppose you had 18 items on a questionnaire that you thought added up to a measure (or index) of racial prejudice. You could check the internal consistency of your index by seeing if respondents answered the odd-numbered questions the same way as the even-numbered questions. If there was not a strong correlation between the levels of prejudice as measured by the two halves, your index would not be very reliable, probably because it was measuring more than one thing.

Split-Plot ANOVA   See *mixed ANOVAs, *mixed designs.

S-PLUS   A programming language that includes statistical and graphics software.

Spread Another term for *dispersion, or the *variability of a group of values or scores.

Spreadsheet   A *computer program that arranges *data and *formulas in a *matrix of *cells. Originally developed for accounting problems, spreadsheets are now popular for organizing many kinds of data.

SPSS   Statistical Package for the Social Sciences. A widely used brand of computer *software that performs most standard statistical analyses of data. Compare *SAS, *Minitab, *LISREL.

Spurious Precision   Said of results reported to more decimal places than is sensible given what is being described. See *rounding error for an example.

Spurious Relation (or Correlation)   (a) A situation in which measures of two or more *variables are statistically related (they *covary), but are not in fact causally linked—often because the statistical relation is caused by a third variable. When the effects of the third variable are removed, they are said to have been *partialed out. See *confound, *lurking variable. (b) A spurious correlation, as defined in (a), is sometimes called an "illusory correlation." In that case "spurious" is then reserved for the special case in which a correlation is not present in the original observations, but is produced by the way the data are handled. Compare *artifact. Also called a "nonsense correlation."

For example, (definition a) if the students in a psychology class who had long hair got higher scores on the midterm than those who had short hair, there would be a correlation between hair length and test scores. However, not many people would believe that there was a causal link and that, for example, students who wished to improve their grades should let their hair grow. The real cause might be sex: Women (who had longer hair) did better on the test. Or that might be a spurious relationship, too. The real cause might be class rank: Seniors did better on the test than sophomores and juniors, and, in this class the women (who also had longer hair) were mostly seniors, whereas the men (with shorter hair) were mostly sophomores and juniors.

SRC   Survey Research Center. Located at the University of Michigan, this organization is best known for its National Election Survey (NES). See *archive, *secondary analysis.

Ss   Abbreviation for the *subjects in a study, as in "Ss were 147 sophomores taking introductory psychology."

SS   Sum of squares, that is, the sum of the squared deviations of scores from the *mean. For an illustration, see *sum of squares.

SSE   *Sum of squared errors. See also *error or *residual.

Stability, Coefficient of   A measure of *reliability calculated by retesting the same subjects with the same test. Also called "test-retest reliability." It is often used when alternate forms of a test are not available. One problem with this *reliability coefficient is determining the appropriate length of time between the test and the retest. If it is too short, subjects may remember items; if it is too long, subjects may change in the interval. See *history effect. Compare *equivalence, coefficient of.

Stack Processing   See *first in-last out (FILO).

Standard Deviation   A *statistic that shows the *spread, *variability, or *dispersion of scores in a *distribution of scores. It is a measure of the average amount the scores in a distribution deviate from the *mean. The more widely the scores are spread out, the larger the standard deviation. The standard deviation is the square root of the *variance. It is symbolized as SD, Sd, s, or lowercase sigma (σ). See *Tchebechev's Theorem and, for an illustration, *sum of squares.

   The standard deviation is an important statistic in its own right and also because it is the basis for many other statistics such as *correlations and *standard errors as well as for all other *standard scores such as the *stanine and the *z-score.

   When the distribution is normal or approximately normal, about two-thirds (68%) of the scores in the distribution will fall between 1 SD below and 1 SD above the mean; about 95% will fall between 2 SDs below and 2 SDs above the mean; over 99% will be located plus or minus 3 SDs from the mean. For example, the height of women aged 18 to 24 in the United States is about 65 inches. The SD is 2.5 inches. Since the distribution is approximately normal, this means that 68% of the women are between 62.5 and 67.5 inches and 95% are between 60 and 70 inches.

   In the following table, three distributions of scores, A, B, and C, are shown with their means and standard deviations. Like other measures of dispersion, the SD tells you how good the measure of *central tendency (in this case the mean) is as an estimate of a value in the distribution. In Distribution A, the mean is a perfect estimate, and the SD is zero. In Distribution C,

Table S.6   Standard Deviation

Distribution											Mean	SD
A	35	35	35	35	35	35	35	35	35	35	35	0.0
B	28	29	30	32	34	36	38	40	41	42	35	5.2
C	1	2	4	5	24	46	65	66	68	69	35	20.6

S

by contrast, the SD is high, and the mean of 35 is a poor estimate of any particular score in the distribution.

Standard Error   Often short for *standard error of the mean or *standard error of estimate. The standard error is a measure of *sampling error; it refers to error in our estimates due to random fluctuations in our samples. The standard error is the *standard deviation of the *sampling distribution of a statistic. It goes down as the number of cases (*n) goes up. The smaller the standard error, the better the *sample statistic is as an estimate of the *population parameter—at least under most conditions.

Standard Error of Estimate (SEE)   The "estimate" is a *regression line. The "error" is how much you are off when using the regression line to predict particular scores. The "standard error" is the *standard deviation of those errors from the regression line. The SEE is thus a measure of the *variability of the errors. It measures the average error over the entire *scatter plot. The lower the SEE, the higher the degree of linear relationship between the two variables in the regression. The larger the SEE, the less confidence one can put in the estimate. Symbolized s_{yx} to distinguish it from s (i.e., the standard deviation of the scores—not the error scores).

Standard Error of Measurement (SEM)   Another way to measure *variability of *errors in predicted scores. Compare *standard error of estimate (SEE). The SEE is smaller than the SEM, which means that the SEM will yield a bigger *confidence interval around the predicted scores.

Standard Error of the Mean   A statistic indicating how greatly the *mean score of a single *sample is likely to differ from the mean score of a *population. It is the *standard deviation of a *sampling distribution of the mean. The standard error of the mean indicates how much the sample mean differs from the *expected value (which is the mean of the sampling distribution of the means). By so doing, it gives an answer to the question: How good an estimate of the population mean is the sample mean? It is calculated by dividing the standard deviation of the sample by the square root of the number of cases (*n). Compare *sampling error.

Standardization of Data   Turning scores or data points into *standard scores. See *z-score, *stanine.

Standardized Measure or Scale   Any statistic that allows comparisons between things measured on different scales or using different *metrics. The best known is a percentage; others include *percentile ranks and *z-scores.

Standardized Regression Coefficient   A statistic that provides a way to compare the relative importance of different variables in a *multiple regression analysis; comparison is possible because the *regression coefficients are expressed as z-scores. Often symbolized as beta and called the *beta

S

weight or *beta coefficient (not to be confused with the "beta" used to symbolize *Type II error). Also called standard partial regression coefficient. (The term "partial" indicates that the effects of other variables have been *held constant.) The unstandardized coefficient *b* is an *asymmetric measure; beta is *symmetric.

Standardized Test   A *norm-referenced test such as the Scholastic Aptitude Test (SAT) or the Graduate Record Examination (GRE). As in any norm-referenced test, an individual's grade is a measure of how well he or she did in comparison to a large group of prior test takers; that prior group was used to determine the norm, i.e., it is the group upon which the test has been "standardized."

"Standardized" also refers to the conditions under which the test is taken (the same directions, materials, amount of time, and so on).

Standard Metropolitan Statistical Area (SMSA)   A U.S. Census category designating an area made up of a central city of 50,000 or more residents and the surrounding region economically tied to it. The term was first used in the 1960 census, when 212 such areas were identified. Compare *census tract.

Standard Normal Deviate   Another term for *standard score or *z-score.

Standard Normal Distribution   A *normal distribution with a *mean of 0 and a *standard deviation of 1.

Standard Partial Regression Coefficient   Another term for *standardized regression coefficient in cases with two or more *independent variables.

Standard Score   Any of several measures of relative standing in a group expressed in *standard deviation units. Using standard scores allows one to compare *raw scores from different *distributions. The most common standard score is the *z-score or "sigma score." See also *T-score, *stanine, *standard deviation. *Proportions and percentages are sometimes called standard scores.

For example, suppose you are a student and you want to compare your scores on the final examinations in each of your five classes. This could be complicated, especially if each of the exams had a different number of questions and each of the classes enrolled a different number of students. You could convert your scores on each of your exams into standard scores. The higher your standard score for each class, the better you did in comparison to other students in that class. The examination with the highest of the five standard scores would be the one on which you did the best.

Stanine   Short for standard ninth. A *standard score scale widely used in school achievement tests. The scale has nine values or stanines: 1, 2, 3, 4,

5, 6, 7, 8, 9. Stanines 2 through 8 are one half of a *standard deviation wide. Stanine 5 straddles the mean (one quarter of a standard deviation above and below). Stanines 1 and 9 include scores that are, respectively, more than 1.75 standard deviations below and above the mean. Stanine distributions have a mean of 5 and a standard deviation of 2. Compare *z-score, *normal curve equivalent.

STATA A statistics and graphics software package.

Statistic A number—such as a *mean or a *correlation coefficient—that describes some characteristic of (the "status" of) a *variable, or of a group of *data.

Strictly speaking, statistics are used to describe *samples, and are usually abbreviated or symbolized by English (Roman) letters. *Parameters are equivalent measures for *populations and are abbreviated or symbolized by Greek letters. In common usage, "statistic" is used for both samples and parameters.

Statistical Conclusion Validity The accuracy of conclusions about *covariation (or *correlation) made on the basis of statistical evidence. More specifically, inferences about whether it is reasonable to conclude that covariation exists—given a particular *alpha level and given the *variances obtained in the study. See *validity, *statistical significance.

Statistical Control Using statistical techniques to isolate or "subtract" *variance in the *dependent variable attributable to variables that are not the subject of study. See *control for, *partial out, *ANCOVA.

Statistical Independence A state in which there is no measured relation between two or more *variables. So called because the statistic describing one variable is independent of (has no association with) a statistic describing another variable or variables. See *orthogonal relationship.

Statistical Inference Using *probability and information about a *sample to draw conclusions ("inferences") about a *population. Specifically, one asks the following question: How likely is it that this result (for example, a mean difference or a correlation) in a sample of this size could have been obtained by chance—if there were no difference in the population from which the sample was drawn? If the probability is low (say less than 5%), then the results are declared statistically significant and are used to draw conclusions about the population. Sometimes called "statistical induction." See *sampling distribution, *external validity, *internal validity.

The same basic statistical concepts are used whether the research is on a representative sample of 1,500 adults surveyed about their attitudes toward political candidates or the research is about memory under three different experimental conditions tested on a group of 90 undergraduates. In the latter case, however, the investigators would put much more emphasis on internal than on external validity, that is, they would be more interested in establishing differences among the treatments than in directly generalizing the findings to a broader population.

Statistical Package   A type of *software that is a collection of *programs for doing statistical procedures with a computer. *SPSS and *SAS are among the more widely used statistical packages in the social and behavioral sciences. See also *BMDP, *LISREL, *Stata, *S-Plus, *MINITAB.

Statistical Power   A gauge of the sensitivity of a *statistical test, that is, its ability to detect effects of a specific size, given the particular *variance and *sample size of a study. It is equal to 1 minus the probability of *Type II or beta error. See also *power of a test, *sensitivity.

Statistical Regression   A tendency for those who score high on any measure to get somewhat lower scores on a subsequent measure of the same thing— or, conversely, for someone who has scored very low on some measure to get a somewhat higher score the next time the same thing is measured. Also called *regression toward the mean, because the second score is likely to move toward or be closer to the mean or average score. Compare *regression, *regression analysis, *regression artifact.

For example, someone who got a 150 on one version of an IQ test would be more likely to get a score of 149 or lower than a score of 151 or higher when taking a second version of the test.

Statistical Significance   Said of a value or measure of a variable when it is ("significantly") larger or smaller than would be expected by chance alone. Compare *level of significance, *probability level, *practical significance, *substantive significance.

It is important to remember that statistical significance does not necessarily imply *substantive or *practical significance. A large sample size very often leads to results that are statistically significant, even when they might be otherwise quite inconsequential.

Calculations of statistical significance are most meaningful on samples that have been randomly selected or on groups that have been randomly assigned. They are of doubtful value without random selection or random assignment. That is because they are designed to measure the probability of a result in a *random* sample.

Statistical Test   Another term for *test statistic, that is, any of several tests of the statistical significance of findings. Statistical tests provide information about how likely it is that sample results are due to *random error.

Statistics   (a) Numerical summaries of data obtained by measurement and computation. (b) The branch of mathematics dealing with the collection and analysis of numerical data. "Statistics" originally meant quantitative information about the government or state—"state-istics." See *official statistics.

Statistics, Descriptive   See *descriptive statistics.

Statistics, Inferential   See *inferential statistics.

Stem-and-Leaf Display   A way of recording the values of a *variable, created by John Tukey, that presents *raw numbers in a visual, *histogram-like display. It is a *histogram in which the bars are built out of numbers.

For example, if 40 students took a final examination in a course, their scores could be shown as in the following stem-and-leaf display. To represent the scores 56, 57, and 59, you put a 5 in the 10s column and a 6, a 7, and a 9 in the 1s column, and so on with the rest of the scores. The display makes it clear that most students scored in the 80s, not many in the 50s and 60s, and so on. The main advantage of a stem-and-leaf display is that it yields a clear picture of the frequency distribution, but unlike most graphic representations, it loses none of the numerical data. Turning the diagram on its side, it is easy to see, for example, that the distribution is *skewed to the left.

Table S.7   Stem-and-Leaf Display of Final Examination Scores

Stem (10's)	Leaves (1's) (Leaf Unit = 1)
5	679
6	02559
7	4466688
8	11233334444567799
9	22236778

Step Diagram (or Step Function)   A graph of a *cumulative frequency distribution. So called because when categorical variables are graphed, the result resembles (usually uneven) stair steps.

Stepdown Selection   Another term for *backward elimination.

Stepup Selection   Another term for *forward selection.

Stepwise Regression   (a) A technique for calculating a *regression equation that instructs a computer to find the "best" equation by entering *independent variables in various combinations and orders. Stepwise regression combines the methods of *backward elimination and *forward selection. The variables are in turn subject first to the inclusion criteria

S

of forward selection and then to the exclusion procedures of backward elimination. Variables are selected and eliminated until there are none left that meet the criteria for removal. Stepwise regression is often, but inconsistently, contrasted with *hierarchical regression analysis, in which the researcher, not the computer program, determines the order of the variables in the regression equation. (b) A less common use of "stepwise" is to describe regression in which the researcher enters the variables in a logical, theoretical order—which is almost the exact opposite of definition (a).

Stochastic   *Random. Thus, a stochastic variable is a *random variable. Said of a *model that is *probabilistic and contains *random elements—as opposed to a *deterministic model. Also used to refer to trial-and-error procedures—in contrast to *algorithmic procedures. Around the year 1900, some statisticians began, for unknown reasons, replacing the word random with the word stochastic. The practice has spread enough that today these synonyms are about equally common in statistical discourse.

Stochastic Models   Attempts to describe the structure of *systematic and *unsystematic error in a set of observations. Compare *ARIMA. Compare *chaos theory.

Stochastic Process   A *random process. When the *probabilities of the occurrence of an event change over time—particularly when the empirical probability approaches the theoretical probability—this is referred to as a stochastic process. See *Markov chain.

Stratified Nonrandom Sample   See *quota sample. Compare *stratified random sampling.

Stratified Random Sample   A random or *probability sample drawn from particular categories (or "strata") of the population being studied. The method works best when the individuals within the strata are highly similar to one another and different from individuals in other strata. Indeed, if the strata were not different from one another, there would be no point in stratifying. The strata have a function similar to blocks in *randomized blocks designs. Stratified random sampling can be proportionate, so that the size of the strata corresponds to the size of the groups in the population. It can also be disproportionate, as in the following example.

Suppose you wanted to compare the attitudes of Protestants, Catholics, Muslims, and Jews in a population in which those four groups were not present in equal numbers. If you drew a *simple random sample, you might not get enough cases from one of the groups to make meaningful comparisons. To avoid this problem you could select random samples of equal size within each of the four religious groups (strata).

Stratified Sampling   Usually short for *stratified random sampling.

Stratifying   Dividing a *population into groups or "strata" before doing research on it. See *stratified random sampling.

Stratum   (plural: strata) A subgroup of a *population. Such groups are used, for example, in *stratified random sampling.

Strength of Association   The degree of relationship between two (or more) variables; often the proportion of the *variability in a *dependent variable explained by or accounted for by the *independent variable(s). *Eta squared, *omega squared, and R^2 are common measures of strength of association. Compare *coefficient of determination, measure of *association. Also called "strength of effect index."

Strength of Effect Index   Another term for a measure of *strength of association. See *effect size.

Structural Coefficients   Another term for unstandardized *path coefficients; they are *regression coefficients in *structural equation models and in path analysis.

Structural Equation   An *equation representing the strength and nature of the hypothesized relations among (the "structure" of) sets of *variables in a *theory or *model. See *structural equation modeling.

Structural Equation Modeling (SEM)   Models made up of more than one structural equation; thus models that describe causal relations among *latent variables and include coefficients for *endogenous variables. SEM is a sophisticated statistical method for testing complex *causal models in which the dependent and independent variables are *latent. A latent variable is one that cannot be observed directly. It is a *construct, that is, a theoretical entity inferred from a pattern of relations (a *structure) among observable variables. SEM combines the techniques of *factor analysis, *path analysis, and *multiple regression analysis, thus allowing researchers to study the effects of *latent variables on each other. Analyses are frequently done with the *LISREL or *EQS computer programs. Also called *analysis of covariance structures. See *confirmatory factor analysis. Compare *canonical correlation analysis.

   For example, take the following apparently simple statement: "The more intelligent a person is, the more likely he or she is to be successful." Both intelligence and success are constructs, latent variables that are constructed out of observations. To begin to study the relationship between intelligence and success, one might sketch in the following kind of *path diagram. (A full SEM path diagram would be much more complex and would include a detailed notation system.) Boxes stand for observed variables; those on the left are measurable aspects of intelligence, those on the right measurable components of success. Circles represent the latent variables.

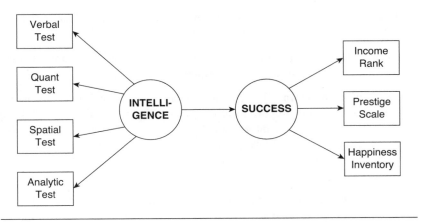

Figure S.7   Structural Equation Model

Structuralism   An approach to anthropology, associated with C. Levi-Strauss, that investigates cultures by studying the structures of language. A main example is the structure of opposites in language-cum-culture, such as sacred-profane and raw-cooked. The structuralism of Levi-Strauss is one of the progenitors of *semiology and has been opposed in recent times by *deconstructionism.

Structure   Any underlying stable pattern among *variables in a system, such as in a social, economic, political, or cognitive system.

Structure Coefficient   A ratio of the *zero-order correlation, r, to the multiple *R in a *regression equation. In short, the structure coefficient for a particular *independent variable is its r with the *dependent variable divided by the R for the whole multiple regression equation. The structure coefficient is widely used in *discriminant analysis to determine the nature of the dimensions on which the groups are differentiated. Also called a *loading, since this is how *factor loadings are calculated in a *factor analysis.

Structured Q Sorts   See *Q sorts, *Q methodology.

Studentized Range Statistic   A t-like statistic used to determine the *critical value in the *Tukey HSD test. See *multiple comparison tests.

Student's *t* Distributions   A family of theoretical *probability distributions used in *hypothesis testing. As with *normal distributions, t distributions are unimodal, symmetrical, and bell-shaped. Called Student's t because the author of the article that made the distribution well known (W. S. Gosset) used the pen name "Student."

The t distribution is especially important for interpreting data gathered on small samples when the population *variance is unknown; when it is known, the z-test is more appropriate. The larger the sample, the more closely the *t* approximates the normal distribution. For samples greater than 120 they are practically equivalent.

Subject   An individual who is studied in order to gather *data for a study. The "individual" may be a person, but it need not be; individual cities, occupations, small businesses, white mice, and so on can also be subjects. The term has increasingly been replaced in recent years by others, including *participant (in psychology), *informant (in anthropology), and *respondent (in survey research). Perhaps the most generic term is *case. Compare *unit of analysis.

Subjective Methods   Any approach to the analysis and evaluation of data based on the researcher's feelings or intuitions about the topic being studied. Often the term is used pejoratively and contrasted with scientific methods. Compare *qualitative.

Subjective Phenomenon   Something that can only be learned from a *subject; hence something that is dependent upon a subject's perceptions and self-reports and not upon information that can be observed directly by someone other than the subject.

For example, the answer to the question, "Do you *like* high-fiber breakfast cereals?" seeks subjective information. On the other hand, "Do you *eat* such cereals?" seeks *objective information; researchers might be able to check the answer to the second question if they doubt the accuracy of the subject's response. In medicine, to take an important example, pain research is based almost entirely on patients' subjective reports of the intensity of their pain.

Subjective Probability   The strength of an individual's belief in the probability of an outcome. *Prior probabilities in *Bayesian inference are subjective probabilities. Also called "personal probabilities."

Subject Matching   See *matched pairs.

Subjects' Rights   See *research ethics, *IRB, *dehoaxing, *anonymity.

Subscript   A number or letter written below and to the right of a symbol to distinguish it from the same symbol with a different subscript.

For example, the number of cases in Group 1 and Group 2 might be written N_1 for Group 1 and N_2 for Group 2.

Subset   A *set contained within another set.

For example, if every *element in one set, accountants (A), is also an element of another set, human beings (H), then A is a subset of H.

Substantive Hypothesis   Like any hypothesis, a substantive hypothesis is a conjecture about the relation between two or more variables. It is called "substantive" because it has not yet been *operationalized and in order to distinguish it from the kind of statistical hypothesis used in *hypothesis testing. See *null hypothesis.

For example, "poor people have fewer 'life chances'," is a substantive hypothesis. To test it, we would need operational definitions of "poor" and "life chances." Perhaps we would use having an annual income less than half of the median for "poor" and average life expectancy for "life chances." We would then need to put the hypothesis into statistical form, perhaps as follows: the mean life expectancy for poor persons (MLP) is less than the mean life expectancy for non-poor persons (MLNP), or MLP < MLNP.

Substantive Significance   Said of a research finding when it reveals something meaningful about the object of study. Often used in contrast with "mere" *statistical significance, which is present when a finding is unlikely to be due to chance alone. Of course, a result that is not statistically significant is often not substantively significant. Also called *practical significance.

For example, suppose we took large random samples of police officers in California and in New York. Comparing some of the data, we find that the mean weight of CA officers is 173 pounds, while that of NY officers is 177 pounds. If the sample were large and representative, even this small difference would be unlikely to be due to chance, or *sampling error, alone; it would be statistically significant. But it would be hard to find someone who thought that the four-pound difference in weight told us anything substantively significant about law enforcement officers in the two states.

Success   In *probability experiments, an event that happens as predicted. The term "success" is usually applied arbitrarily to one of two ways a *Bernoulli trial can turn out. Success might be drawing a black card; "*failure" would then be drawing a red card.

Sufficient Condition   In *causal analysis, a *variable which, by itself, is always enough ("sufficient") to bring about a change in another variable. There are few, if any, sufficient conditions in the social and behavioral sciences. Compare *necessary condition.

Summated Scale   A *scale or *index made up of several items measuring the same variable. The responses are given numbers in such a way that responses can be added up ("summated").

For example, say we have an attitude index about government responsibility made up of five questions in the following form: "The government in Washington should see to it that women and minorities do not experience job discrimination; strongly agree = 4; agree = 3; disagree = 2; strongly disagree = 1." Respondents' answers to the five questions can be added to arrive at a single number that measures the positive or negative strength of their attitudes. On a five-item index using the above numbers, the highest possible score would be 20—strongly agree (4) with all five statements: $4 \times 5 = 20$; the lowest would be 5—strongly disagree (1) with all five: $1 \times 5 = 5$.

Summative Evaluation   *Evaluation research conducted in the latter stages of a program to assess its impact or to determine how well it has met its

S

goals. Summative evaluations are often undertaken to help policymakers decide whether to continue, expand, reduce, or terminate a program's funding. Compare *formative evaluation.

When assessing products, formative evaluation would be part of product development; one would find summative evaluations in *Consumer Reports* magazine.

Sum of Squared Errors (SSE) In a *regression analysis, the SSE are what you are trying to minimize when you use the *ordinary least squares criterion. The *errors in question are the vertical distances of the observed scores from the *regression line (or predicted scores). Also called the sum of squared *residuals.

Sum of Squares (SS) The result of adding together the squares of *deviation scores. *Analysis of variance is in fact analysis of sums of squares. See *within, *between, *total, and *regression sum of squares. Not to be confused, as it often is when doing calculations, with the "square of sums," that is, all the scores first added together to get a sum, which is then squared.

For example, the following table lists the scores on a test, calculates the *mean, subtracts the mean from each score, squares each of those results, and adds (sums) these numbers. Computing the sum of squares is a step on the way to calculating the *variance and the *standard deviation. The variance is found by dividing the sum of squares by the number of scores minus 1 (238 divided by 6 = 39.67 in this example), and the standard deviation is calculated by taking the square root of the variance (which equals 6.3 in this example).

Table S.8   Sum of Squares

Scores	Minus Mean		Deviation Score	Deviation Score Squared
88	−80	=	8	64
86	−80	=	6	36
84	−80	=	4	16
80	−80	=	0	0
77	−80	=	−3	9
73	−80	=	−7	49
72	−80	=	−8	64
560				238
560/7 = 80 (mean)				(sum of squares)

Superfluous Variable In *regression analysis, an *independent or *predictor variable that adds nothing to the total variance explained (*R^2) by the regression equation.

Superscript A number or letter written above and to the right of a symbol to denote its *power.

For example, the 3 in 8^3 is a superscript and means 8 times 8 times 8 (which equals 512).

Suppressor Effect   See *suppressor variable.

Suppressor Variable   A variable that conceals or reduces (suppresses) a relationship between other variables. It may be an *independent variable unrelated to the *dependent variable but correlated with one or more of the other independent variables. Removing the suppressor variable from the study raises the correlation between the remaining independent variable(s) and the dependent variable. Compare *confound.

For example, suppose we were testing candidates for the job of forest ranger. We're sure that a good forest ranger must know a lot of botany; we also think that verbal ability has no effect on rangers' job performance. We give a written botany test to help us pick good rangers. But since the test is written, candidates with stronger verbal ability will tend to get higher scores, even if their knowledge of botany is no greater. In this example, verbal ability is the suppressor variable. It gets in the way of studying what we are interested in: knowledge of botany and rangers' job competencies.

To take another example, education tends to increase people's liberalism. But education also tends to increase people's incomes, and increased income tends to reduce people's liberalism. In this case, income is the suppressor variable. See the following illustration and compare it with that for *reinforcer effect.

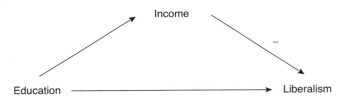

Figure S.8   Suppressor Variable

Surrogate Variable   Another term for *proxy variable. See *social indicator.

Survey   A research *design in which a *sample of subjects is drawn from a *population and studied (often interviewed) in order to make inferences about the population. A *census is a survey of an entire population; when contrasted with a census, a survey is then called a *sample survey. See *random sample.

Survival Analysis   The study of how long subjects persist ("survive") in a state, a categorical variable. By contrast, *time-series analysis studies change in a continuous variable over time. Survival analysis has been used in medical research to study the duration of illnesses, in demography to study life expectancy, and in organizational studies to analyze the survival of small businesses. Survival analysis is a variety of *event history analysis in which there are a limited number of states or conditions. Analytic techniques used in survival analysis are forms of *regression analysis. See *competing risks for an example.

Syllogism   A pattern of formal *deductive argument that contains a major premise, a minor premise, and a conclusion. If (but only if) the two premises are true, the conclusion necessarily (logically) follows.
For example:

Major Premise: All undergraduates like statistics.

Minor Premise: Mary is an undergraduate.

Conclusion: Mary likes statistics.

Symbolic Interactionism   An approach in sociology to the study of how people interact in groups that grew up in the context of *pragmatism in the 1930s. Most closely associated with the work of Hubert Blumer, it stresses close observation of how individuals in interaction use symbols to accomplish their purposes.

Symbolic Logic   A branch of logic that uses formal symbols, rather than ordinary language, to express its concepts. The purpose is to avoid the ambiguities of ordinary, informal, natural language. Compare *formal theory.

Symmetric Measure   A *statistic that has the same value regardless of which *variable is thought of as *dependent and which is *independent. See *asymmetric measure. Examples include *Pearson's r and *Kendall's tau.

Synchronic   Said of research focusing on events that occurred at the same time. Usually contrasted with *diachronic. Compare *cross-sectional study.

Synergism   A term used to describe the increase in an effect when two treatments are administered jointly. A term for an *interaction effect, used mostly in the context of drug testing, especially when the interaction produces a good result.

Synthesis   Combining parts (such as *data, *concepts, or *theories) to make a new whole. Sometimes contrasted with *analysis, which involves disassembling wholes to study their parts. In practice, analysis is often a step on the way to synthesis.
For example, suppose we wished to calculate an average (*mean) income figure for citizens of all South American nations. Since nations report aggregate national income figures somewhat differently (some

include transfer payments and others do not), we might have to *disaggregate the figures. We could then analyze the parts that went into making up the whole reported by each nation. Then, we could perhaps figure out a way to synthesize them into a new, more general report about all the nations.

SYSTAT   A statistical software package.

Systematic Error   Measurement error that is consistent, not *random. Another term for *bias. Compare *random error. Also called "invalidity." See *validity, *ARIMA.

For example, if you were to sample a population in an area code by dialing phone numbers at random, your sample would be biased; it would systematically overrepresent people with more than one phone number (home, office, cell) and exclude people who did not have a phone.

Systematic Review   Often another term for *meta-analysis. Systematic review can also refer to *literature reviews (whether using quantitative summaries or not) that stress clearly enunciated and replicable procedures for selecting studies to be reviewed and for drawing conclusions from them—as opposed to a more impressionistic approach.

Systematic Sample   A sample obtained by taking every "*n*th" subject or case from a list containing the total *population (or *sampling frame). The size of the *n* is calculated by dividing the desired sample size into the population size. Also called a "list sample."

For example, if you wanted to draw a systematic sample of 1,000 individuals from a telephone directory containing 100,000 names, you would divide 1,000 into 100,000 to get 100; hence, you would select every 100th name from the directory. You would start with a randomly selected number between 1 and 100, say 47, and then select the 147th name, the 247th, the 347th, and so on.

It is always possible to do a simple random sample in cases where you can do a systematic sample (both require a complete list of the population). Simple random sampling is a more trustworthy method, but it is usually less convenient.

Systematic Variance   Variance due to a cause that always influences values in one direction. Compare *bias, *cause, *systematic error.

Systems Theory   An approach or perspective in several disciplines that emphasizes studying the interrelations of the parts of a whole (the system) more than studying components in isolation from their position in an organized whole.

Tails (of a Distribution) Values at the high and low end of a graphic display of a distribution. If these values are extreme at one end, the distribution is *skewed. The distribution represented by the following graphic is right skewed because it has a long skinny tail on the right (high) end of the distribution. The left (low) end of the distribution is more "fat-tailed." See *two-tailed test of significance.

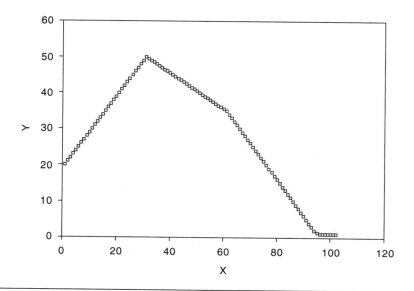

Figure T.1   Tails of a Distribution

Target Group (or Population)   The group about which a researcher wishes to draw conclusions; another term for a *population about which one "aims" to make inferences. Also called the "parent population."

Tau [T, τ]   (a) *Kendall's tau. (b) *Goodman and Kruskal's tau. (c) Symbol for the *population total, not the sample total. (d) A common symbol for "time" in a *time series.

Tautology   A statement or conclusion that involves *circular reasoning. This might involve, for example, taking the presence of an effect to argue for the presence of a cause, which is then claimed to bring about the effect. To avoid tautologies, one needs independent measures of causes and effects.

Taylor-Russell Tables   Lists of calculations of predictions based on the *proportions of individuals who will be successful on a given *outcome, such as college graduation. The calculations involve using different cutoff scores on predictive tests, such as the SATs. These tables were important in the early history of *decision theory and are still used to illustrate some of its concepts.

Taxonomy   (a) A system of classification, originally of plants or animals; the term is sometimes used more broadly in the social and behavioral sciences to refer to classifications of personality types, political systems, and so on. (b) The systematic study of general principles of classification.

Tchebechev's Theorem (or Inequality)   A method for calculating the maximum *probability of a particular score in a *distribution of scores. It is used to set rough (and *conservative) outer limits for how many measurements will fall within 1, 2, and 3 *standard deviations from the mean, regardless of the shape of the distribution.

For example, according to the theorem, at least 3/4 of the measurements in any distribution will fall in the range from −2 to +2 standard deviations from the mean.

t Distribution   See *Student's t distribution.

T Distribution   The distribution used in *Hotelling's T-test. See *T-test.

Term   Part of an *equation. For example, in a *regression equation, the *error term, e, is added to the other values in the equation.

Test of Normality   Comparison of a set of observations to see whether they could have been produced by *random sampling from a *normal *population. A test of normality involves comparing the sample distribution with a normal distribution.

Test-Retest Reliability   A *correlation between scores on two administrations of a test to the same subjects. A high correlation indicates high

T

reliability. Subtracting the correlation from 1 $(1 - r)$ gives you an estimate of *random error. See *stability, coefficient of.

Test Statistics (a) Statistics used to test a finding for *statistical significance—not to describe a *sample or to estimate a *population parameter. They test hypotheses about sample statistics, but provide no information about the samples. The *t-test, *chi-squared statistic, and *F ratio are examples of test statistics. The test statistic is the *calculated value* for the t, F, or chi-squared, not the distribution. These statistics (t, F, and chi-squared) have known distributions when the *null hypothesis is true, which means that they can be used to see if the sample distribution conforms to the null hypothesis. (b) Statistics about tests and test items, such as statistical aspects of their *validity and *reliability. See *standardized test, *Rasch models, *item response theory.

Tetrachoric Correlation A correlation between two *continuous variables that have been *dichotomized. Symbolized r_{tet}. It gives an estimate of what *Pearson's r would have been, had the variables not been collapsed.

Before the availability of computers to do the grunt work, r_{tet} was frequently used for quick estimates of Pearson's r. It remains useful today mainly for data containing many gaps and errors, such as imperfect historical records. Compare the *phi coefficient, which is used for dichotomous, not dichotom*ized,* variables.

Theorem (a) A statement or formula in mathematics or logic deduced from other statements or formulas. A theorem is derived from premises; it is not an *assumption. (b) An idea accepted or proposed as a clear truth, often as part of a general theory. Compare *axiom, *postulate.

Theoretical Probability Distribution See *probability distribution, definition (b).

Theory A statement or group of statements about how some part of the world works—frequently explaining relations among phenomena. While theory is usually distinguished from practice in ordinary language, Kurt Lewin's oft-quoted phrase nicely captures the belief of most researchers: "There is nothing so practical as a good theory." Compare *hypothesis, *law, *realism.

Theory Trimming In *path analysis, deleting paths whose *coefficients are not *statistically significant.

Thick Description Used in *ethnography to refer to highly detailed, specific descriptions of cultural life. The term, coined by Gilbert Ryle, is most closely associated with the anthropology of Clifford Geertz, who specifically contrasted it with such "thin" descriptions as those implied by *behaviorism.

Threats to Validity   Problems that can lead to false conclusions. The term was introduced by Campbell and Stanley to refer to the characteristics of various research methods and designs that can lead to spurious or misleading conclusions.

Discussions of threats to validity often lead researchers to recommend using more than one method (see *mixed-method research, *triangulation). Because different kinds of research designs are open to different kinds of threats, you can reduce the *risk of error by using two or more methods. Researchers sometimes use a list of threats to validity as a checklist to review before putting the finishing touches on a design.

Threshold Effect   An effect in a *dependent variable (DV) that does not occur until a certain level (threshold) has been reached in an *independent variable (IV). There are two general types: (1) when increases in the IV past the threshold continue to produce increases in the DV. For example, inflation may occur only when productivity drops below a certain rate, but further drops continue to increase inflation; (2) when additional amounts of the IV do not lead to increases in the DV—often because the DV exists in only one of two states (is *dichotomous). For example, a smoke alarm will not ring when there is very little smoke in the air; but once the threshold is met and it does ring, more smoke will not make it ring louder.

Three-Way ANOVA   *Analysis of variance with three *independent variables. Four interactions are possible in a three-way ANOVA, whereas in a two-way ANOVA only one is.

Thurstone Scaling   (a) A method of *scale construction in which judges assign weights or degrees of intensity to prospective scale items. The extent to which judges agree determines whether particular items are included in the scale. The method is seldom used today, largely because it is slow, expensive, and no more *reliable than *Likert scales. (b) A method of assessing the difficulty of items on a test pioneered by Thurstone; it employs paired comparisons between groups taking the test, and it is still widely used.

Time Series   A set of measures of a single *variable recorded periodically, over time, such as the annual rainfall in Miami from 1900 to the present. See *smoothing and *moving average.

Time-Series Analysis   (a) Analysis of changes in variables over time; *multiple regression is sometimes used for this purpose, but *ARIMA models are more often employed. (b) Any of several statistical procedures used to tell whether a change in *time-series data is due to some variable that occurred at the same time or was due to coincidence. See *interrupted time-series analysis.

Time-Series Data   Any data arranged in chronological order.
For example, the annual suicide rate in the United States from 1900 to the present would be time-series data, as would the Dow Jones daily averages for the last 18 months, as would the weekly spelling test scores of a class of 6th graders. Compare *survival analysis.

Time-Series Design   See *interrupted time-series design, *time-series analysis.

TLA   Three-letter acronym. Many fields, including research methods and statistics, make extensive use of TLAs.

Tobit Analysis   A technique in *regression analysis used when the *dependent variable may be either zero or any positive number. For this kind of dependent variable, the *ordinary least squares criterion cannot be applied. Also known as the "censored regression model."
For example, say the research question is, "How much money will individuals spend on microcomputers this year?" The answer will fit into one of two categories: For most individuals the answer will be zero; for the rest it will range widely from a few hundred to several thousand dollars.

Tolerance   (a) Support for the equal rights and liberties of others, usually others one finds threatening. (b) An allowable margin of error in measurement. (c) In *multiple regression analysis, the tolerance is the proportion of the variability in one *independent variable (IV) not explained by the other IVs in the regression equation. The bigger the tolerance, the more useful the IV is to the analysis; the smaller the tolerance, the higher the *collinearity. Compare *variance inflation factor. (d) In some psychological studies, tolerance can mean the ability to withstand a potentially harmful *treatment.

Total Determination, Coefficient of   In *regression analysis, the proportion of the total variance in the *dependent variable explained by the *independent variables. Symbolized R^2. Also called coefficient of *determination.

Trait   An enduring characteristic; it could be a physical characteristic such as sex or a psychological trait, such as shyness. Compare *attitude, which in comparison to a trait is learned and always refers to something outside a person—that is, an attitude is always toward something.

Trait-Treatment Interaction   A *research design using *factorial ANOVA to determine whether there is an *interaction effect between *traits and *treatments; for example, whether a management effectiveness seminar has better results with women or men. Also called "attribute-treatment interaction."

Transformation   Changing all the values of a variable by using some mathematical operation. One common example is changing proportions into percents by multiplying them by 100. Another widespread practice is to use a transformation to reduce the complexity of a table reporting large numbers, as when one reports income in thousands of dollars. In that case the transformation is done by dividing each number by 1,000. Taking the *log (or the square root) of the values in a distribution is done to facilitate analysis, often by making distributions conform to statistical *assumptions. See *linear and *nonlinear transformations for more examples; also *logistic and *probit regression.

Transformed Standard Scores   *Transformations performed on *z-scores, usually in order to eliminate decimals and negative numbers. For examples see *Z-score and *normal curve equivalent. The Educational Testing Service transforms z-scores by multiplying them by 100 and adding 500 to arrive at the familiar range from 200 to 800 used to report scores on the Scholastic Aptitude Test and the Graduate Record Examination.

Transpose   A *matrix formed by interchanging the rows and columns (vectors) of another matrix. See *vector for an example.

Treatment   (a) In *experiments, a treatment is what researchers do to subjects in the *experimental group but not to those in the *control group. A treatment is thus an *independent variable. (b) Used broadly to mean almost any *predictor variable. For example, in a study of the effect on the traffic accident rate of changing the speed limit, the speed limit would be the "treatment." See *level, *condition.

Treatments-by-Subjects ANOVA   *Repeated-measures ANOVA.

Treatment Group   Another term for *experimental group, especially in the context of a *clinical trial.

Tree Diagram   A way of depicting a series of possible events using "branches" to illustrate different outcomes. It is used in *probability calculations and *decision theory. See *decision tree.

   For example, the following diagram shows the six possible outcomes, and their *probabilities, of a two-out-of-three-set tennis match between A and B. A's record shows that she wins .70 (70%) of the sets she plays; B's set winning rate is 30%. The most likely outcome is that A will win in two sets (.49). The least likely is that B will win the match after having lost a set (.063).

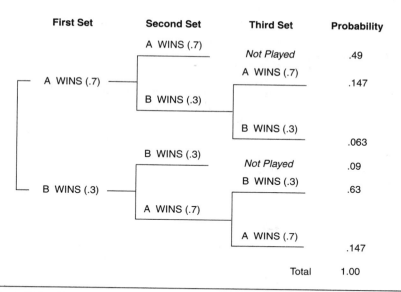

First Set	Second Set	Third Set	Probability
	A WINS (.7)	Not Played	.49
A WINS (.7)		A WINS (.7)	.147
	B WINS (.3)		
		B WINS (.3)	.063
	B WINS (.3)	Not Played	.09
B WINS (.3)		B WINS (.3)	.63
	A WINS (.7)		
		A WINS (.7)	.147
		Total	1.00

Figure T.2   Tree Diagram

Trend   Movement in one direction of the values of a *variable over a period of time. See *secular trend and, for an example, *moving average. Compare *fluctuation.

Trend Analysis   A form of *regression analysis used to discover *nonlinear relations in *time-series data.

Trend Line   A line depicting a *trend. See *moving average and *ARIMA.

Trend Study   An investigation of a variable over time in which the population studied does not remain the same.

For example, a study of college graduates' starting salaries in their first jobs from 1980 to the present would study a different group of graduates each year. Compare *cohort study, *panel study.

Triangular Distribution   A *probability distribution of a *discrete variable that forms a triangle. The illustration at *underlying distribution is an example.

Triangulation   Using more than one method to study the same thing. The term is loosely borrowed from trigonometry, where it refers to a method for calculating the distance to a point by looking at it from two other points. See *mixed-method research, *multimethod research.

For example, if you were interested in people's attitudes toward environmental issues, you could look at patterns of voting behaviors on environmental candidates and issues; or you could interview leaders of the

Sierra Club, the Nature Conservancy, and similar groups; or you could conduct a survey of a representative sample of the entire population. Or, you could do all three and put the results together, in which case you could say that you had used a research strategy of triangulation.

Trimmed Mean   A measure of *central tendency that allows the researcher to deal separately with a *distribution's *outliers. It is a *mean computed without the extreme observations. Compare *truncated sample.

For example, an 80% trimmed mean would be the mean calculated using only the central 80% of the values in the distribution; the high and low 10% would be eliminated (trimmed). Compare *interquartile range.

Trough   In a *time series, a point lower than the preceding and following points. Compare *spike.

True Experiment   An *experiment. "True" is contrasted with the methods of *quasi-experiments and *natural experiments. The key distinction is that, unlike in other research designs, in a true experiment subjects are *randomly assigned to *treatment groups and the researchers manipulate the *independent variable(s).

True Score   In measurement theory, a score is thought to consist of the true score plus or minus random measurement error. If the errors are random, they will cancel one another out in the long run, after many measurements, and yield the true score. That means that the true score can be assumed to be the mean of a large number of measurements.

Truncated Distribution or Sample   A distribution with data missing, nonexistent, or cut off at one end, such as income distribution that omits anyone below or above a certain amount. Sometimes truncated and censored are used as synonyms. When a distinction is made, truncation usually refers to an incomplete *sample,* while censored refers to incomplete measurements of a variable. Compare *censored data or samples, *tobit analysis.

Trustworthiness   The equivalent of *validity when referring to *qualitative research. The equivalency is conceptual. In practice, the means of assessing validity or trustworthiness vary between qualitative and quantitative approaches. Because of the special nature of qualitative data, standard quantitative techniques (*validity coefficients, for example) are inappropriate. On the other hand, many questions of the two research traditions are the same—for example, have researchers accurately recorded what they have been told by respondents or informants?

T-Score   (uppercase T) Another term for either (a) the *Z-score or (b) the *normal curve equivalent (NCE). Note: Not to be confused with the *t statistic or the *t-test.

The T-score (definition a) indicates deviation from a *mean. The mean is scored 50 and one *standard deviation is scored 10. Thus, a T-score of

70 is 2 standard deviations above the mean; 40 is one standard deviation below.

TSS   Total *sum of squares

t **Statistic**   The number that is tested in a *t-test, that is, the number that is compared to the *critical region. Compare *F ratio, *test statistic.

t-**Test**   A test of *statistical significance, often of the difference between two group *means, such as the average score on a manual dexterity test of those who have and have not been given caffeine. Also used as a *test statistic for *correlation and *regression coefficients. See *Student's *t* distribution. Compare *z*-test.

A *two-tailed *t*-test is used to test the significance of a "nondirectional" hypothesis, that is, an hypothesis that says there is a difference between two averages without saying which of the two is bigger. A *one-tailed *t*-test is called "directional" because it tests the hypothesis that the mean of one of the two group averages is bigger. When the researcher is uncertain about which is larger, the two-tailed test should be used.

There are several formulas for *t*. The one to use depends upon the nature of the data and the groups being studied, most often upon whether the groups are *independent or *correlated.

T-**Test**   The uppercase *T* has been overused. It can mean (a) *Hotelling's *t*-test (b) a rank-ordered test of trends in a time series, and (c) a variant of the *Mann-Whitney *U* test. All these are quite distinct from the *T-Score and the *t-test (lowercase *t*).

Tukey Line   A *regression line based on *medians (rather than *means, as in the *least squares criterion). It is more *resistant to *outliers than the ordinary least squares (OLS) regression line.

Tukey's Honestly Significant Difference (HSD) Test   After conducting an *analysis of variance of the differences in group means, the researcher knows whether *some* group means are significantly different from the overall mean. To determine *which* means are significantly different, Tukey's HSD test can be used. See *post hoc comparison, *multiple comparisons test.

For example, suppose 150 overweight subjects were randomly assigned (30 each) to five different weight-loss programs to see if there were any significant differences among the programs. At the end of 10 weeks, the average weight loss for each group was computed, and an ANOVA *F ratio showed that the differences among the groups were statistically significant. However, to determine which groups contributed most to those differences, that is, which means were significantly different from others, individual comparison tests such as Tukey's HSD would have to be conducted.

Two-Bend Transformation   A transformation for relations between two variables which, when plotted on a graph, produces a line with two bends. The most common transformations of this type are the *logistic and the *probit.

Two-by-Two Design   Said of a research design with two *independent variables, each with two values. See *N-by-M design.

The following two-by-two *factorial table shows the percentage of students passing a pass/fail examination given to male and female college sophomores and seniors.

Table T.1   Two-by-Two Design: Percentage Passing Examination

	Sophomores	*Seniors*
Men	40	62
Women	48	71

Two-Sided Test   See *two-tailed test.

Two-Tailed Test of Significance   A statistical test in which the *critical region (*region of rejection of the *null hypothesis) is divided into two areas at the tails of the *sampling distribution. Also called a "nondirectional test," "two-sided test," and "two-direction test." A two-tailed test is more demanding, or more conservative, than a *one-tailed test. See *Type I error, and, for an example, *t-test.

Two-Way ANOVA   *Analysis of variance with two *independent variables. It is used to study the effects of two independent variables separately (their *main effects) and together (their *interaction effect). Compare *one-way ANOVA.

Type I Error   An error made by wrongly rejecting a true *null hypothesis. This might involve incorrectly concluding that two variables are related when they or not, or wrongly deciding that a sample statistic exceeds the value that would be expected by chance. Also called *alpha error. See *hypothesis testing, *false positive.

Type II Error   An error made by wrongly retaining (or accepting or failing to reject) a false *null hypothesis. Also called *beta error. See *Type I error, *false negative.

Type I and Type II errors are inversely related; the smaller the risk of one, the greater the risk of the other. The probability of making a Type I error can be precisely computed in advance for a particular investigation, but the exact probability of Type II error is generally unknown.

U   (a) Symbol for the *Mann-Whitney *U* test statistic. (b) Symbol for the *disturbance term in a *regression model. (c) In *path analysis, the symbol for unanalyzed effects or unmeasured variables. It is composed both of the effects of *variables not included in the *model and the disturbance or *random error. (d) In *factor analysis, the symbol for "unique factor," that is, the part not explained by the *common factors.

Unbalanced Designs   Said of *factorial designs when there are unequal numbers of observations for different *factors, or when the *cells contain unequal numbers of subjects. Also called "nonorthogonal factorial designs."

Unbiased   Fair. Said of a research *design that is free from any characteristic that would misrepresent the evidence or of a method of analysis that does not distort the findings.

Unbiased Error   An error that is *random in the sense that, in the long run, positive and negative unbiased errors cancel one another out and sum to zero.

Unbiased Estimate   A *sample statistic that is free from *systematic error leading it to over- or underestimate the corresponding *population parameter.

Unbiased Sample   A sample that is free of any *systematic error in the way it was defined or in the way data were collected for it. See *random sample, *non-response bias. Note the distinction between an unbiased sample and an *unbiased estimate. An unbiased estimate will not repair the damage done by a biased sample.

U

Unbiased Sample Variance   A method of computing the *sample variance so that it is an *unbiased estimate of the *population variance, usually by dividing the *sum of squares by $n - 1$ rather than n. Statistical packages often use $n - 1$ by default even when it is not appropriate because the data are population, not sample, data (the package has no way of knowing).

Uncertainty   A state in which outcomes are unknown and in which there is no basis for calculating the *probabilities of outcomes. This contrasts with a situation characterized by *risk, in which the probabilities of at least some of the outcomes can be estimated. Sometimes the terms risk and uncertainty are used interchangeably.

Underidentified Model   A *regression model or equation containing too many unknowns to be solved. See *identification problem.

Underlying Distribution   The distribution of all possible outcomes of an *event.

For example, if the event is the result (sum) of a roll of two dice, there are 11 possible sums: 2, 3, 4 . . . 12, distributed as in the following table. The table shows that there is one way to roll a 2, 2 ways to roll a three, 6 ways to roll a 7, and so on. This is an example of a *triangular distribution.

Table U.1   Underlying Distribution of the Sum of a Roll of Two Dice

```
            7
            6                                 _*_
Number of   5                         _*_           _*_
   Ways      4                  _*_                        _*_
            3            _*_                                      _*_
            2      _*_                                                  _*_
            1 _*_                                                            _*_
            0
               1    2    3    4    5    6    7    8    9   10   11   12

            Sum of the Dice
```

Underspecified Model   A causal model in which important variables have been omitted. See *specification problem, *path diagram.

Unimodal Distribution   A distribution with only one *mode. Compare *bimodal.

Union   In *set theory, a *set formed by including all of the *elements of two other sets. Symbolized $A \cup B$.

Unit A synonym for 1 in some statistical discussions. To say, for example, that the *standard deviation of a *normal distribution is "unit," means that it is 1.

Units of Analysis The persons or things being studied. Units of analysis in research in the social and behavioral sciences are often individual persons but may be groups, political parties, newspaper editorials, unions, hospitals, schools, monkeys, rats, reaction times, perceptions, and so on. A particular unit of analysis from which data are gathered is called a *case.

Unity What statisticians sometimes like to call 1. For example, "this value approaches unity" means that it gets close to 1.

Univariate Analysis (a) Studying the distribution of *cases of one variable only—for example, studying the ages of welfare recipients, but not relating that variable to their sex, ethnicity, and so on. Compare *multivariate analysis. (b) Occasionally used in *regression analysis to mean a problem in which there is only one *dependent variable—a usage that conflicts with the more common meaning in definition (a).

Universal Constant See *E, definition (c).

Universal Set All things to be considered in any one discussion. Usually symbolized S. In *sampling theory, the universal set is the *population.

Universe Another term for *population, one that is more likely to be used in logic than in statistics.

UNIX A widely used *operating system.

Unobtrusive Methods Procedures for collecting data, often about social life in natural settings, that are hidden or at least that do not interfere much with what is being studied. Measuring the popularity of museum exhibits by observing patterns of wear in the flooring is a classic example. Observing people in a public place, such as a coffee shop, while pretending to be an ordinary patron is another example, but one that raises ethical questions about *informed consent.

Unsaturated Model See *saturated model.

Unstandardized Score A score in the original *metric or units of measurement, one that has not been transformed into a *z-score or other *standard score.

U-Shaped Distribution A *frequency distribution in which the largest frequencies are at the extremes of the scores; this results in a graphic that resembles the letter U. The term "inverted U-shaped distribution" is sometimes used to describe frequency distributions in which the most frequent values are toward the middle and the least frequent are at the high and low ends of the scores. Sometimes called a "goalpost distribution."

The graphic indicates the racial composition of high schools in Cook County, Illinois (Chicago and suburbs). The x axis indicates the percentage of students attending the high schools who are Black. The y axis gives the number of high schools. Almost all of the high schools are less than 10% Black or more than 90%. Very few are in the middle range of 20% to 80%.

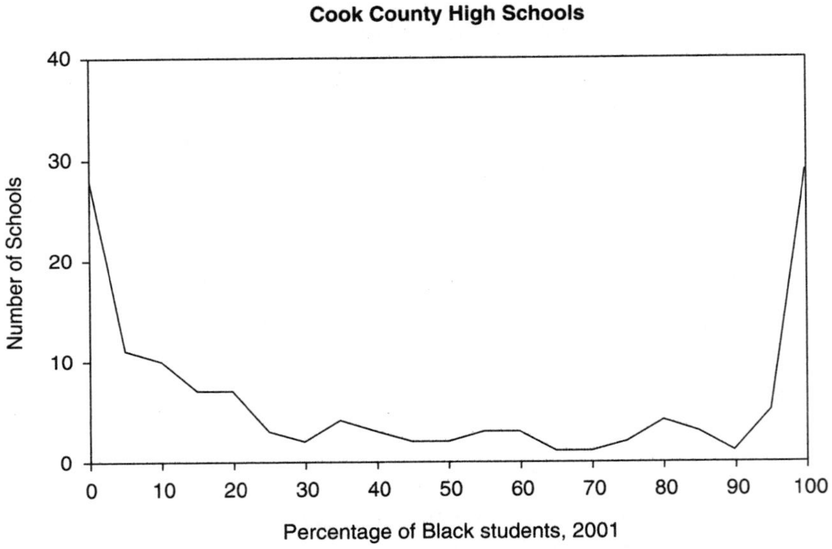

Cook County High Schools

Figure U.1   U-Shaped Distribution

Utility   The usefulness or satisfaction derived from an outcome. Utility is subjective, but it is usually given a number. The satisfaction gotten from, say, a new pair of shoes may differ from one person to the next. Someone might say "on a scale of 1 to 10, I'd give these shoes a 7," while you might have given them only a 4. In economics, utility is a determinant of the price a person will pay. In *decision theory, utility is combined (as a weighting factor) with the *probability or *expected value of an outcome to make a choice. Compare *minimax and *maximin strategy.

V See *Cramer's *V.*

Validity A term to describe a measurement instrument or test that accurately measures what it is supposed to measure; the extent to which a measure is free of *systematic error. Validity also refers to *designs that help researchers gather data appropriate for answering their questions. Validity requires *reliability, but the reverse is not true.

    For example, say we want to measure individuals' heights. If all we had was a bathroom scale, we could ask our individuals to step on the scale and record the results. Even if the measurements were highly reliable, that is, consistent from one weighing to the next, they would not be very valid. The weights would not be completely useless, however, since there generally is some *correlation between height and weight.

    Social scientists have not found it easy to define validity. To try to explain it they have created several subcategories, including: *internal, *external, *content, *construct, *criterion-related, and *face validity.

Validity Coefficient A *Pearson correlation between two sets of scores, such as scores on an old and a new version of a test, or between a *predictor and a *criterion variable.

Value Added The difference between inputs and outputs. In manufacturing, that would mean the costs minus the selling price. In education it might mean the difference between students at the beginning and at the end of a year of schooling.

Value-Free Said of science when researchers keep their personal values out of the collection and interpretation of evidence. As Max Weber pointed out long ago, value-free means that researchers place the values of science above their other values, not that they have no values. Contrast *value-laden.

V

Value Judgment   A belief or statement that something is good (or bad). All value judgments are opinions, but not all opinions are value judgments. Value judgments are usually contrasted with statements of fact, but in practice the distinction is often unclear. See *objective.

Value-Laden   Said of any research or theory that contains (is weighed down by) the values of the researcher or theorist. Often contrasted with *objective and *value-free.

Variability   The *spread or *dispersion of scores in a group of scores; the tendency of each score to be unlike the others. More formally, the extent to which scores in a *distribution deviate from a *central tendency of the distribution, such as the *mean. The *standard deviation and the *variance are two of the most commonly used measures of variability. For an example, see standard deviation.

Variable   *noun* (a) Loosely, anything studied by a researcher. (b) Any finding that can change, that can vary, that can be expressed as more than one value or in various values or categories. The opposite of a variable is a constant. (c) In algebra, a variable is an unknown. See *categorical, *continuous, *dependent, *independent, and *random variables.

Examples of variables include anything that can be measured or assigned a number, such as unemployment rate, religious affiliation, experimental *treatment, grade point average, and so on. Much of social science is aimed at discovering and explaining how differences in some variables are related to differences in others.

Variable Parameter Model   A model in which the *population *parameters being estimated are assumed to be variable, not constant.

Variance   A measure of the spread of scores in a *distribution of scores, that is, a measure of *dispersion. The larger the variance, the farther the individual cases are from the *mean. The smaller the variance, the closer the individual scores are to the mean.

Specifically, the variance is the mean of the sum of the squared deviations from the mean score divided by the number of scores. (See *sum of squares for an example.) Taking the square root of the variance gives you the *standard deviation.

Variance Components Analysis   Another term for *random effects model or *Model II ANOVA.

Variance-Covariance Matrix   A square matrix in which the entries on the diagonal are *variances and the others are *covariances of the variables measured. Widely used in *factor analysis and *structural equation modeling. Compare *correlation matrix (a correlation is a *standardized covariance).

Variance Explained   In *regression analysis, the variance in the *dependent variable that is associated with or accounted for by the independent variables. R^2 is a common measure of the variance explained.

Variance Inflation Factor (VIF)   A measure of *collinearity in multiple *regression analysis. This statistic is 1 divided by the *tolerance. Therefore low tolerances result in high VIFs and vice versa. The lowest possible VIF is 1.0 when there is no collinearity. If the tolerance were .2, the VIF would be 5.0 (1.0/.2 = 5.0).

Variance of Estimate   A measure of the *variability of the points around a *regression line or surface. The square root of the variance of estimate is the *standard error of the estimate, that is, the standard deviation of the *residuals. The variance of estimate is also called the *mean square residual (MSR).

Variate   (a) Another term for *variable. (b) Another term for *random variable. (c) Any specific value of a *variable, as 37 would be a variate of the variable "age." (d) A composite variable in *canonical correlation, similar to a *factor in factor analysis. See *multivariate analysis, *univariate analysis, *covariate.

Variation   Commonly used to mean the extent of the "variety" in a *variable, such as the spread of values in a *distribution of the values for a particular variable. It is also often used to refer loosely to any measure of *dispersion.

   More formally, but less commonly, the variation is a step on the way to calculating other measures of dispersion, such as the *variance and the *standard deviation. This variation is the total of the squared deviations from the mean score (that is, the *sum of squares). Since adding up all the squared deviations from the mean can produce a huge sum, the number is divided by the number of cases (minus 1 for a sample) to get the variance. To get the standard deviation, one takes the square root of the variance.

Variation, Coefficient of   A measure used to compare the dispersion or variation in groups of scores. To obtain the coefficient of variation one divides the *distribution's *standard deviation by its *mean. It is appropriate only for scores measured on a ratio scale. Compare *Gini coefficient.

Variation Ratio   A measure of *dispersion that can be used with *categorical data. It is the *relative frequency of the nonmodal scores or values.

   For example, if the relative frequency of the modal score were .40 (as it is in the example at *frequency distribution), then the variation ratio would be .60.

Varimax   A widely used method of *orthogonal rotation of the axes in a *factor analysis. As its name suggests, it maximizes the variances of the factors.

Vector   (a) In *matrix algebra, a column (or a row) of a *matrix, or a matrix with only one column or one row. A vector is a set of ordered values; thus, it has both size and direction. Any particular value in a vector (e.g., 7 in the following example) is called an "element" of the vector. The number of elements in the vector is its "dimension," which in the example is 5, that is, there are 5 elements in Vector **b**.

(b) In *MANOVA, a vector is a mathematical expression representing a subject's score on all the *dependent variables; the mean of these individual vectors is the *centroid.

Table V.1   Vector

Column Vector		
$b = \begin{bmatrix} 3 \\ 7 \\ 9 \\ 4 \\ 8 \end{bmatrix}$	row vector $b' = [3\ 7\ 9\ 4\ 8]$	

NOTE: **b'** is the *transpose of **b**.

Venn Diagram   A type of graph using circles to represent variables and their relationships. The rectangle represents the *universal set (*population), and the circles inside the rectangle are sets (variables). *Union and *intersection are represented by the circles' overlap.

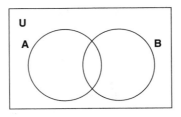

Figure V.1   Venn Diagram

The rectangle U in Figure V.1 represents the "universe" of all the adults in a particular city. The circle A stands for all of the registered Republicans, and the circle B stands for the city's adult African Americans. The overlap of A and B indicates those African Americans who are registered members of the Republican Party.

V

Verisimilitude   (a) The state of being close to or similar to the truth; competing scientific theories are sometimes thought of as having different degrees of verisimilitude. (b) Used in qualitative methods to mean a narrative that appears to be true and/or a text that adheres to the norms of a writing genre.

Verstehen   A German word meaning "understanding." It was used by Max Weber and Wilhelm Dilthey to refer to a method of interpreting social interaction that involves putting oneself in the place of another. Broadly today, any method using empathetic means of gaining insight into the motives or behavior of another, such as *role-playing. See *Geisteswissenschaften.*

Vicious Circle   A definition or an argument that assumes what it is trying to prove. Compare *tautology.

VIF   *Variance inflation factor.

Vital Statistics   Statistics about births, deaths, marriages, and divorces—so called because they have to do with life (*la vita*). Also used more loosely to include information about health and diseases.

Volatility   Another term for *standard deviation, especially when used to describe stock market prices.

Volunteer Bias   Any of several problems that arise in drawing valid conclusions from research because participation in the research is voluntary. Since people cannot ethically or legally be compelled to be research subjects, volunteer bias is always a potential source of inaccuracies in generalizing from research samples to broader populations. See *research ethics, *sampling error.

   For example, if you are studying authoritarianism in college students and if volunteers for studies tend to be less authoritarian than those who refuse to participate, your sample will not be representative.

W   Symbol for (a) the *Wilcoxon test of statistical significance and for (b) *Kendall's coefficient of concordance and for (c) the *Shapiro-Wilks test.

Wald-Wolfowitz Test   A *nonparametric test of the *null hypothesis that two *samples have been drawn from identical *populations. Sometimes called the "runs test."

Wald Statistic   A *test statistic use for assessing the *statistical significance of individual *coefficients in a *logistic regression.

Wave   In a *panel or longitudinal study, when the same subjects are interviewed more than once, each session is called a wave.

Wavelet   A formula used to represent *time-series data. Several varieties of wavelet exist.

Weighted Average (or Mean)   A procedure for combining the means of two or more groups of different sizes; it takes the sizes of the groups into account when computing the overall or grand mean.

For example, say 239 students in Economics 201 took the midterm and you know the average grade of each of three groups: economics majors, economics minors, and others, as summarized in the following table. But you want to know the grand mean of all 239 students. You cannot just add up the three groups' mean grades and divide by 3 (which would give you an average of 83.2). So many more majors than minors and others took the test that you have to give their mean more weight to get the true overall average. What you do is, for each group multiply the mean (Column B) by the number of people in the group (Column A), add up the results and divide by the total number of students who took the test: 21,228.4/239 = 88.8.

Table W.1   Computing a Weighted Average

Group	(A) Number	(B) Mean Grade	(C) A × B
Majors	168	93.2	15,657.6
Minors	51	78.8	4,018.8
Others	20	77.6	1,552.0
Total	239		21,228.4

Weighted Data   (a) Any information given different weights in calculations, as when the final examination counts twice as much as (is weighted double) the midterm. (b) Data whose values have been adjusted to reflect differences in the number of *population units that each *case represents. See disproportionate *stratified sampling, *weighted mean.

For example, suppose we want to generalize about the attitudes of all 80-year-olds. We have a *sample of 100 men and 100 women, all 80 years old. We might want to give the women's attitudes more weight, since there are many more 80-year-old women than men, and the goal is to generalize to the attitudes of all 80-year-olds.

Weighted Least Squares   A version of *ordinary least squares regression in which all the variables are multiplied by a particular number or "weight." This is sometimes done to correct for *heteroscedasticity. Compare *generalized least squares. See *loess.

Weighted Moving Average   See *moving average, *smoothing.

Weighted Sample   In survey research, a sample in which *weighting has been applied to different categories (such as ethnicity) to make the sample more representative of the population. For example, if a group was known to comprise 10% of the population and the sample contained only 8% of members of that group, researchers could weight that group's responses more heavily when computing statistics for the sample as a whole. The actual numbers used to adjust the figures are called "sample weights." See *weighted data.

Weighting   (a) *verb* Multiplying a variable by some amount in order to make it comparable to other variables. See *weighted average. (b) *noun* The number by which the variable is multiplied.

Welch Test   A test of statistical significance used to test the difference between *means when the data violate the assumption of *homogeneity of variances required of other tests, such as the *F and *t-tests.

White Noise   Completely *random variation, that is, variation containing no *systematic elements. The term is used often to describe randomness in *time-series data. See *noise.

W

Widespread Biserial Correlation   A measure of *association, which, like the ordinary *biserial correlation, is used when one *variable is *continuous and the other has been dichotomized by the researcher. In the widespread version, the researcher is interested in extremes in the dichotomized variable, but not in the middle range of scores.

Wilcoxon Test   More fully, the Wilcoxon "signed-rank" or "rank-sum" test for *ordinal data. A *nonparametric test of *statistical significance for use with two *correlated samples, such as the same subjects on a before-and-after measure. See the *Mann-Whitney U test and the *Kruskal-Wallis test, which require *independent samples.

Wilcoxon-Pratt Test   A modification of the *Wilcoxon test used to deal with tied ranks.

Wilks's Lambda   A widely used *test statistic for equality of group means (*centroids) in a *MANOVA and in other multivariate tests. The test yields an *F ratio, which is then interpreted in the same way as any other F ratio.

Within-Group Difference (or Variance or SS)   In an *ANOVA, the part of the total *variance attributable to differences among the subjects in a (*control or *experimental) group. The within-group variance is compared to the between-group variance. If the within-group variance is small compared to the between-group variance, this indicates that the *treatments the groups received were significantly different. See F ratio, which is a ratio of the between-group differences to the within-group differences. Also known as *error variance and "interclass variance."

Within-Group ANOVA   Another term for *repeated-measures ANOVA.

Within-Samples Sum of Squares   Another term for *error sum of squares.

Within-Subjects ANOVA   Another term for *repeated-measures ANOVA.

Within-Subjects Design   A before-and-after study or a study of the same subjects given different *treatments. A research design that *pretests and *posttests within the same group of subjects, that is, one which uses no *control group. Compare *between-subjects design, *repeated measures design.

Within-Subjects Variable (or Factor)   An *independent variable or factor for which each subject is measured more than once—at different *levels or *conditions. Compare *between-subjects variable. See *counterbalancing for an example.

X The letter commonly used to symbolize an *independent, *predictor, or *explanatory variable. When discussing more than one such variable, they are usually called X_1, X_2, X_3, and so on.

X^2 *Chi-squared.

x Axis The horizontal axis on a graph. Also called the *abscissa.

X-Bar A letter X with a line over the top of it ($\bar{x}$); this is an often-used symbol for the *mean of a *sample. The mean score for a *population is usually symbolized by the Greek letter, *mu.

X Variable The *independent or *predictor variable (or cause), so called because, when graphed, it is plotted on the *x axis (also called the abscissa or horizontal axis).

Y Letter commonly used to symbolize the *dependent, *outcome, or *criterion variable.

Y′ Called "Y prime," the symbol for the predicted value of Y, the *dependent variable, in a *regression equation. Also Y-hat, ($\hat{Y}$), and $Y*$.

Yates's Correction (for Continuity) An adjustment in the computation of the *chi-squared statistic to improve its accuracy for 2 × 2 tables, especially those with small cell values. "Continuity" refers to the fact that the correction is made because one is using a continuous distribution (chi-squared) to estimate a discrete distribution.

   Yates's continuity correction is less widely used than it once was, largely because many statisticians think that it may overcorrect for the possibility of *Type I error and thus increase the chances of *Type II error. Compare *Fisher's exact test.

y Axis The vertical axis on a graph. Also called the *ordinate. The *dependent variable is usually plotted on the y axis.

Yield Curve A line on a graph that results when the interest rate for bonds or other securities is plotted against the length of time they have to be held before they pay off. These curves generally slope upward because the longer investors must hold investments, the higher the rates they want.

Y Intercept The point where a *regression line intersects the *y axis. Sometimes referred to simply as the *intercept. In a *regression equation, it is usually symbolized by the letter a. The intercept is the value of the dependent (Y) variable when all the independent (X) variables are zero. See the graphic at *intercept.

Yule's *Q*   A measure of *association for two *dichotomous variables. It is a simplified version of *gamma for 2 × 2 tables. It is a *PRE measure.

Y Variable   The *dependent or *outcome variable (or effect), so called because, when graphed, it is plotted on the *y axis (also called the *ordinate or vertical axis).

Z   Uppercase *Z*, see *Fisher's *Z*. Lowercase *z*, see *z-score.

z **Axis**   The vertical axis in a representation of a three-dimensional space.

Zero-Order Analysis   Analysis of original data. If you use that analysis to do another analysis, it would be a first-order analysis. If you used the first-order analysis to do another, that would be second-order, and so on.

Zero-Order Correlation   A correlation between two *variables in which no additional variables have been *controlled for (or held constant or partialed out). A *first-order correlation is one in which one variable has been controlled for; a second-order correlation controls for two, and so on. Zero-order associations are the opposite of those measured by *multivariate analysis.

Zero-Sum Game   Any game (or, more broadly, social situation) in which one player (or social actor) can gain only at the expense of another and in which one player gains exactly as much as another loses.

For example, if you and I were to play poker for money, it would be a zero-sum game; the only way you could win money would be for me to lose it, and vice versa. The term "zero-sum" comes from the fact that if you add my losses to your winnings, the result, or sum, is zero. Certain forms of social interaction (such as poker) are clearly zero-sum games; others (such as donating blood) may benefit both the giver and the receiver; and for others (such as affirmative action) the extent to which they are zero-sum situations is open to debate.

z-**Score**   (lowercase *z*) The most commonly used *standard score. It is a measure of relative location in a *distribution; it gives, in *standard deviation units, the distance from the mean of a particular score. In *z*-score notation,

Z

the mean is 0 and a standard deviation is 1. Thus, a z-score of 1.25 is one and one quarter standard deviations above the mean; a z-score of −2.0 is 2 standard deviations below the mean. Therefore, z-scores are especially useful for comparing performance on several measures, each with a different mean and standard deviation.

For example, say you took two midterm exams. On the first you got 90 right, on the second 60. If you knew the *means and *standard deviations, you could compute z-scores for each of your exams to see which one you did better on. The procedure is to take your score, subtract from it the mean of all the scores, and divide the result by the standard deviation (in symbols, $z = X - M/Sd$), where X is your score, M is the mean, and Sd is the standard deviation. The following table shows how to compute your z-scores and compare them. In this example, you did better (ranked higher in the class) on your second midterm. Your score of 60 was 2 standard deviations above the mean; your 90 was only 1 standard deviation above.

Table Z.1   Computing Your z-Score

First Midterm	Second Midterm
$X = 90$	$X = 60$
$M = 80$	$M = 42$
$Sd = 10$	$Sd = 9$
$(90 - 80)/10 =$	$(60 - 42)/9 =$
$10/10 = 1$	$18/9 = 2$

Z-score   (uppercase Z) A *standard score in which the mean of the distribution is 50 and the *standard deviation is 10. The Z-score is obtained by transforming the z-score (multiplying z by 10 and adding 50), which is why the Z-score is called a *transformed standard score. The main advantage of this transformation is that it eliminates decimals and negative numbers. Sometimes, but not consistently, called the *T score.

z-Test   A test of *statistical significance based on the assumption that a *sample distribution approximates a *standard normal distribution. It is less frequently used than the t-test, because the t-test is more accurate with small samples, but there is little difference between the two when samples are large (greater than 30).

z-Transformation   Another term for *r-to-z transformation.

z Value   A *z-score (lowercase z) used as a *test statistic when the *sample distribution is assumed to equal or approximate a *normal distribution.

Suggestions for Further Reading

R eaders wishing to consult statistics and methodology texts to supplement this dictionary have *many* works from which to choose. Listed below are works that are very clearly written, and/or are widely cited accounts, and/or are representative of some of the many disciplines that use the methods defined in this dictionary.

Suggested books are grouped by level of difficulty. The more elementary books in Group I are basic enough that most readers can read them without help. Group II contains more advanced volumes that, while usually starting with the basics, move quickly to more difficult topics and examine them in greater depth. Group III lists other dictionaries and reference works that contain methodological and statistical terms.

Special mention should be made of the Sage series *Quantitative Applications in the Social Sciences*. Works in this series introduce readers to advanced topics in short booklets. By focusing on concepts rather than proofs and derivations, the booklets can cover advanced topics without requiring of readers more than basic statistical knowledge.

I. Elementary Methodology and Statistics

Babbie, E. (1995). *The practice of social research* (7th ed.). Belmont, CA: Wadsworth.

Black, T. R. (1993). *Evaluating social science research: An introduction.* Thousand Oaks, CA: Sage.

Creswell, J. W. (2002). *Educational research: Planning, conducting, and evaluating quantitative and qualitative research.* Upper Saddle River, NJ: Merrill/Prentice Hall.

Creswell, J. W. (2003). *Research design: Qualitative, quantitative, and mixed methods approaches.* (2nd ed.). Thousand Oaks, CA: Sage.

Freedman, D., Pisani, R., & Purves, R. (1998). *Statistics* (3rd ed.). New York: W.W. Norton.

Gall, M. D., Borg, W. R., & Gall, J. P. (1996). *Educational Research: An introduction.* (6th ed.). New York: Longman.

Huck, S. K. (2004). *Reading statistics and research* (4th ed.). Boston: Allyn & Bacon.

Jaeger, R. M. (1990). *Statistics: A spectator sport.* (2nd ed.). Thousand Oaks, CA: Sage.

Moore, D. S. (2001). *Statistics: Concepts and controversies* (5th ed.). New York: W. H. Freeman.

Neuman, W. L. (1997). *Social research methods: Qualitative and quantitative approaches* (3rd ed.). Boston: Allyn & Bacon.

Norusis, M. J. (1990). *SPSS introductory statistics student guide.* Chicago: SPSS.

Ramsey, F. L., & Schafer, D. W. (2002). *The statistical sleuth: A course in methods of data analysis* (2nd ed.). Pacific Grove, CA: Duxbury.

Sirkin, R. M. (1995). *Statistics for the social sciences.* Thousand Oaks, CA: Sage.

Utts, J. M. & Heckard, R. F. (2004). *Mind on statistics* (2nd ed.). Belmont, CA: Thomson.

II. More Advanced Works on Methodology and Statistics

Agresti, A. & Finlay, B. (1997). *Statistical methods for the social sciences* (3rd ed.). Upper Saddle River, NJ: Prentice Hall.

Chernick, M. R., & Friis, R. H. (2003). *Introductory biostatistics for the health sciences: Modern applications including bootstrap.* Hoboken, NJ: Wiley.

Cohen, J., Cohen, P., West, S. G., & Aiken, L. S. (2003). *Applied multiple regression/correlation analysis for the behavioral sciences* (3rd ed.). Mahwah, NJ: Erlbaum.

Fisher, R. A. (1970). *Statistical methods for research workers* (14th ed.). New York: Holt, Rinehart & Winston.

Fisher, R. A. (1971). *The design of experiments* (9th ed.). New York: Hafner.

Fox, J. (1997). *Applied regression analysis, linear models, and related methods.* Thousand Oaks, CA: Sage.

Glass, G. V., & Hopkins, K. D. (1996). *Statistical methods in education and psychology* (3rd ed.). Boston: Allyn & Bacon.

Greene, W. H. (2000). *Econometric analysis* (4th ed.). Upper Saddle River, NJ: Prentice Hall.

Hays, W. L. (1988). *Statistics* (4th ed.). Fort Worth, TX: Holt Rinehart, & Winston.

Johnson, R. A., & Wichern, D. W. (2002). *Applied multivariate statistical analysis* (5th ed.). Upper Saddle River, NJ: Prentice Hall.

Keppel, G., & Zedeck, S. (1989). *Data analysis for research designs.* New York: Freeman.

Kline, R. B. (2004). *The principles and practice of structural equation modeling* (2nd ed.). New York: Guilford.

Knoke, D., & Bohrnstedt, G. W. (1994). *Statistics for social data analysis* (3rd ed.). Itasca, IL: Peacock.

Myers, J. L., & Well, A. D. (2003). *Research design and statistical analysis* (2nd ed.). Mahwah, NJ: Erlbaum.

Mohr, L. B. (1990). *Understanding significance testing.* Thousand Oaks, CA: Sage.

Newton, R. R., & Rudestam, K. E. (1999). *Your statistical consultant: Answers to your data analysis questions.* Thousand Oaks, CA: Sage.

Norusis, M. J. (1988). *SPSS advanced statistics user's guide.* Chicago: SPSS.

Pedhazur, E. J. (1997). *Multiple regression in behavioral research: Explanation and prediction* (3rd ed.). Fort Worth, TX: Harcourt Brace.

Pedhazur, E. J., & Schmelkin, L. P. (1991). *Measurement, design, and analysis: An integrated approach.* Hillsdale, NJ: Erlbaum.

Shadish, W. R., Cook, T. D., & Campbell, D. T. (2002). *Experimental and quasi-experimental designs for generalized causal inference.* Boston: Houghton Mifflin.

Tabachnick, B. G., & Fidell, L. S. (2001). *Using multivariate statistics* (4th ed.). Boston: Allyn & Bacon.

Tukey, J. W. (1977). *Exploratory data analysis.* Reading, MA: Addison-Wesley.

III. Dictionaries and Reference Works

Ammer, C., & Ammer, D. S. (1984). *Dictionary of business and economics.* New York: Free Press.

Andrews, F. M., Klem, L., Davidson, T. N., O'Malley, P., & Rodgers, W. (1981). *A guide for selecting statistical techniques for analyzing social science data* (2nd ed.). Ann Arbor, MI: Institute for Social Research.

Bannock, G., Baxter, R. E., & Davis, E. (1992). *The Penguin dictionary of economics* (5th ed.). London: Penguin.

Borowski, E. J., & Borwein, J. M. (1991). *The Harper Collins dictionary of mathematics.* New York: Harper Collins.

Calhoun, C. (2002). *Dictionary of the social sciences.* Oxford: Oxford University Press.

Cramer, D., & Howitt, D. (2004). *The Sage dictionary of statistics.* London: Sage.

Dodge, Y. (2003). *The Oxford dictionary of statistical terms.* Oxford: Oxford University Press.

Everitt, B. S. (2002). *The Cambridge dictionary of statistics* (2nd ed.). Cambridge: Cambridge University Press.

Lewis-Beck, M. S., Bryman, A., & Liao, T. F. (2004). *The Sage encyclopedia of social science research methods* (3 vols.). Thousand Oaks, CA: Sage.

Porkess, R. (1991). *The HarperCollins dictionary of statistics.* New York: HarperCollins.

Schwandt, T. A. (1997). *Qualitative inquiry: A dictionary of terms.* Thousand Oaks, CA: Sage.

Scriven, M. (1991). *Evaluation thesaurus* (4th ed.). Thousand Oaks, CA: Sage.

Upton, G., & Cook, I. (2002). *A dictionary of statistics.* Oxford: Oxford University Press.

Weisstein, E. W. (2003). *CRC concise encyclopedia of mathematics* (2nd ed.). London: CRC.

About the Author

W. Paul Vogt is Professor of Research Methods and Evaluation at Illinois State University, where he has won both teaching and research awards. He specializes in the evaluation of educational programs and is particularly interested in integrating multiple methods in program evaluation. His other books include: *Tolerance & Education: Learning to Live With Diversity and Difference, Quantitative Research Methods for Professionals,* and *Education Programs for Improving Intergroup Relations* (co-edited with Walter Stephan).